Rational Ecology

To my parents,
Henryk and Lorna Dryzek

Rational Ecology

Environment and Political Economy

John S. Dryzek

Basil Blackwell

Copyright © Basil Blackwell Inc. 1987

First published 1987

Basil Blackwell Ltd
108 Cowley Road, Oxford, OX4 1JF, UK

Basil Blackwell Inc.
432 Park Avenue South, Suite 1503
New York, NY 10016, USA

Library of Congress Cataloging in Publication Data

Dryzek, John S., 1953–
  Rational ecology.

  Bibliography: p.
  Includes index.
  1. Environmental policy—Decision making.
  2. Social choice.
  I. Title
  HC79.E5D79   1987   363.7'0561   87–6587

  ISBN 0–631–15574–0

British Library Cataloguing in Publication Data

Dryzek, John
  Rational ecology: the political economy of
  environmental choice.
  1. Social choice 2. Political science
  I. Title
  320.01'9   JA77

  ISBN 0–631–15574–0

Typeset in Sabon on 10/12 pt
by Alan Sutton Publishing Ltd
Printed in the USA

Contents

vi   *Contents*

Preface

The means that societies employ to make collective choices have all kinds of ramifications. This study rests on the contention (justified more fully in chapter 1) that these "social choice mechanisms" strongly affect the kind of world that exists or develops. To provide a purchase on larger questions of political economy, I focus on the capacity of these mechanisms to handle ecological problems. Ecology is used here as a kind of "window" on social choice. Other windows might be provided by social justice, individual liberties, economic efficiency, and so forth. I hope that the view through an ecological window will engage those whose interest in political economy stems from different concerns. Impatient readers whose main interest is political economy rather than ecology may be forgiven for skipping the last section of chapter 3, the first two sections of chapter 4, and chapter 5.

Why focus on ecology? During recent years human beings have, collectively, made great strides in their capabilities to eradicate both themselves and their fellow-creatures from the planet. Catastrophes stemming from chronic environmental degradation and resource exhaustion, though, remain live possibilities rather than near-certainties. However, such problems eventually need to be confronted, be it in a matter of years, decades, or centuries. So how may ecological limits be recognized, respected – and perhaps evaded? Are our current means for reaching collective decisions and solving social problems up to the task? To answer these questions, I focus on the major existing forms of social choice: markets, polyarchy, law, and the like (part II). Given some rather severe flaws in these established mechanisms, I explore some innovative alternatives (part III). Both established and innovative structures are analyzed according to a standard I call "ecological rationality" (part I).

Social choice mechanisms exist in the eye of the beholder. There is no well-defined set of mechanisms "out there" waiting to be discovered – unlike, for example, the set of countries of the world. Arguably, every instance of collective human choice – whether the 1983 British general election, a Politburo decision to purchase grain from the USA, the Moroccan market in snowmobiles, or the Franco-Prussian War – has its own unique structure.

My study will be sensitive to and will draw upon particular cases, but its primary focus is upon specific *types* of choice. To generate analyses, evaluations, and designs of any power, one must group mechanisms into categories. Taxonomy is a prerequisite to explication, explanation, and evaluation. The categories into which I group social choice mechanisms are defined by "ideal types," which real-world instances of choice can be expected to approximate to varying degree, or to mix in varying proportion.

One way of classifying forms of collective choice is through reference to the distinctive kind of *coordination* across actors that a mechanism embodies. (Coordination is what makes these mechanisms social rather than individual.) This classification is developed (and defended) in chapter 6. Reflecting this classification, each of the chapters of part II investigates the capabilities of one (or more) established kinds of social choice. My analysis of less familiar and more innovative forms in part III follows a "logic of search" in institutional design, with its roots in the best that established mechanisms have to offer.

This inquiry revolves around two key turning points, in chapters 4 and 14, respectively. The first moves from a contemplation of the peculiar nature of ecological problems to a set of criteria for evaluating social institutions. The stage is thereby set for a scrutiny of the world's major existing social choice mechanisms. The second is a transition from an evaluation of these existing forms to the design of less familiar innovative structures. This second fulcrum develops the idea of an "open society" as both a paragon of the ecological capabilities of extant social choice mechanisms and a foil for an agenda of institutional innovation.

In all cases, my method for dealing with a mechanism (established or innovative) is to:

(a)   make as strong a case as possible *for* the ecological rationality of the mechanism;
(b)   make as strong a case as possible *against* that mechanism in terms of the same standard;
(c)   synthesize an overall judgment from these two arguments.

If I reach more negative than positive judgments, I hope it is not through taking cheap shots at straw men or degenerate forms. If I condemn a mechanism, I hope it is because that mechanism *at its strongest* still fails.

My use of the term "social choice" may jar the sensibilities of those accustomed to a narrower usage. Social choice has in some circles come to connote an area of study at the intersection of political science and economics concerned with the aggregation of conflicting, predetermined, individual preferences to reach group outcomes. My broader use interprets social choice mechanisms as problem-solving rather than merely preference-aggregating devices. This usage bears some resemblance to Lindblom's (1977) notion of "political–economic systems."

The astute reader may discern in the pages that follow an element of schizophrenia in my treatment of rationality in individual and collective decision. On the one hand, a central theme is that we need to move beyond dominant instrumental–analytic conceptions of rationality in social choice, under which reason is viewed solely as the capacity to devise, select, and effect good means to clarified ends. On the other, I make full use of the apparatus of instrumental–analytic rationality in my own dissections and discussions of different kinds of social choice. Let me suggest in my defense that one can use instrumental analysis to criticize instrumental–analytic rationality in exactly the same way as one can talk in English about the deficiencies of the English language. (I owe this parallel to Feyerabend, 1978: 81.) Looser parallels may be found in Taoism and Zen. As Lao Tzu puts it in the *Tao Te Ching*, "the Tao that can be expressed is not the eternal Tao" – but it can be expressed only in the language it transcends. Zen Buddhists make full use of the intellect in their quest for transcendence of reason and direct experience beyond intellectual apprehension. One makes use of the tools one has. If instrumental–analytic rationality perishes at its own hand, then surely its death is more convincing than if it were at the hands of some other kind of worldview.

With a few exceptions, the kinds of questions raised in this book have been downplayed by mainstream political scientists and economists in recent years. A number of metaphors spring to mind: Nero fiddling with Rome burning, or a rearrangement of deckchairs on the Titanic. The Titanic metaphor can be extended in the most interesting directions. Many ecologists are aware of icebergs in the vicinity, and seek to convince us that the ship of state should chart a course to avoid them. Most economists would be more concerned with ensuring a utility-maximizing arrangement of deckchairs. Most political scientists would worry about whether their methods for analyzing the voting behavior of the people in the deckchairs were scientific. The trouble is that the iceberg-avoidance task is left to those with scant knowledge of political–economic systems, who consequently produce naive, sweeping, and erroneous analyses. Icebergs merit more serious attention.

Acknowledgments

... when I had thus put mine ends together,
I shew'd them others, that I might see whether
They would condemn them, or them justifie:
And some said, let them live; some, let them die:
Some said, *John,* print it; others said, Not so:
Some said, It might do good; others said, No.

John Bunyan, *Pilgrim's Progress*
("The Author's Apology")

Of those who said, "John, print it," Dave Bobrow, Peggy Clark, Brian Cook, Bob Goodin, Welling Hall, Charles Lindblom, Aaron Wildavsky, and Oran Young variously provided inspiration, advice, encouragement, or criticism, for all of which I am grateful.

Part I

Foundations

1

Prelude

AN ECOLOGICAL PROBLEM: ACID RAIN

A new term entered the environmental lexicon in the 1970s: "acid rain." Acid rain itself is as old as the industrial revolution – if not older. But for the first time, it was widely recognized as a distinct kind of pollution. Rumblings of discontent began in Sweden, with the realization that a number of lakes had become too acidic to support fish. Norway reported a similar story. Soon it became apparent that Scandinavia did not suffer alone. Reports of damaged forests and acidified lakes came in from north-eastern North America (especially Ontario, New England, and the Adirondack Mountains). Mountain lakes in Colorado revealed a similar story. Reduced crop yields, corroded buildings, and diminished quality of drinking water were attributed to acid rain in a variety of locations. Public concern peaked in West Germany in the mid-1980s, where government studies confirmed *Waldersterben* (literally, the death of the forests), especially in Bavaria and the Black Forest. Half the nation's trees seemed to be damaged, and some observers raised the specter of a treeless Germany. Germany's European neighbors were less worried. The British Forestry Commission released a survey in 1985 showing negligible acid-related damage to forests in the United Kingdom, although studies of particular regions (such as the south-west Highlands in Scotland) have revealed the presence of highly acidic surface water. In France, the Office National des Forêts at first claimed there was little pollution-related damage to French trees; but in 1984 it had admitted substantial damage in particular regions (especially the Vosges Massif and Alsace). Sporadic evidence from eastern Europe suggested that acid rain fell on communists as well as capitalists. It was discovered in the Third World too, especially in and around industrializing nations such as Brazil and Korea.

The environmental degradation associated with acid deposition has caused substantial political fallout. Least surprising is the fact that countries and regions importing acid rain (notably Norway, Sweden, Canada, New England, and Colorado) have taken the political offensive against net exporters (UK, France, USA, Ohio Valley, and Arizona). The exporters generally downplay any problems they are causing. Some see crocodile tears and ulterior motives in the importers' complaints; for example, it is sometimes claimed that Canadian attempts to restrict US acidic emissions are no more than a cover for Canada's desire to improve the competitive edge of its electrical power industry in US markets close to its border. The US federal government is not particularly sympathetic to the Canadian case; its tone changes, however, when it comes to the acid rain the US imports from Mexican copper smelters.

Domestic political fallout has been most extensive in West Germany, where the primary electoral beneficiary has been the ecologically minded Green Party. Christian Democrats, Social Democrats, trade unions, and businesses have shown more concern with economic growth than forest growth. Outside Germany, acid rain tends to prompt interregional or international political friction, rather than domestic dispute. Thus, Swedes direct their ire at the UK, not at each other.

The causes and consequences of acid rain remain shrouded in no little mystery. It is clear that acid rain (and associated phenomena such as acid snow, mist, and dry deposition) is a product of atmospheric chemical transformations undergone by sulfur dioxide and oxides of nitrogen. Sulfur dioxide has both natural sources (e.g., volcanos) and man-made sources. The guilty human sources are mostly coal-burning power plants and ore smelters. The more prominent culprits include power stations in the American Midwest, the United Kingdom, and the Ruhr; copper smelters in Mexico and Arizona; and power plants in Poland which burn brown coal. Oxides of nitrogen have more diverse sources, including automobile exhausts.

The rest of the acid rain story is more uncertain. It is difficult, if not impossible, to trace a particular acid rainfall to a specific source (or sources). In part this is because pollution is transported and dispersed across hundreds of kilometers. More important is the fact that the various pollutants act and interact in variable ways while they are in the atmosphere, depending on meteorological conditions. Further controversy persists over how much acidity is natural and how much is induced by pollution. And it is by no means clear whether rainfall is becoming more or less acidic with time in particular locations (the required long-term trend data do not exist). Additional ambiguity surfaces when it comes to the effects of acid rain on ecosystems. It is not easy to trace specific cases of ecological damage (such as stunted tree growth) to acid rain. The mechanisms through which acid affects plant life are not well understood (though leaching of toxic metals from soil is suspected of playing a major role). Observed damage can often be linked to other plausible causes, such as drought, frost, or localized pollution. Effects that do occur vary dramatically with soil types, kinds of vegetation, topography, and micro-climate. Therefore there is huge local variation in damage severity and kind.

Ambiguity over the causes and consequences of acid rain elicits a variety of reactions. It prompts some investigators to engage in systematic ecological research, such as the development of large computer models for the explanation and prediction of the meteorology and atmospheric chemistry of acid rain.[1] Ambiguity also facilitates symbolic policy responses in the form of finance for further research – for example, through the Interagency Task Force created by the US 1980 Acid Deposition Act. Scientific uncertainty too is adduced as an argument to eschew any attack on acid rain – at least until more and better data become available. Coincidentally, those who equivocate on this basis (such as the executive branch of the US federal government) often have an economic or political stake in the relevant polluting activities. Those who continue to suffer the effects of acid rain can only grin and bear their suffering; for as long as responsibility for pollution damage cannot be established, traditional legal remedies are unavailable.

Technical solutions *are* available (if sometimes expensive) for some aspects of the acid rain problem, especially those associated with sulfur dioxide. Alternatives include the "scrubbing" of emissions by passing them through a spray of lime and water (which removes up to 95 percent of the sulfur dioxide), "fluidized bed" coal combustion, and coal "washing" prior to burning. Newly constructed coal-burning power plants in the Netherlands, USA and West Germany are required to scrub their emissions. Some of these methods produce a further pollutant in the form of sulfurous sludge, but the Japanese have developed methods for extracting valuable sulfur, sulfuric acid, and gypsum from the scrubbings. Technical solutions to oxides of nitrogen pollution are more elusive. At the other end of the chain of effect, sporadic and variably efficacious attempts have been made to neutralize the effects of acid by "liming" lakes.

Despite the existence of these technical solutions and substantial political pressure from people concerned about or suffering from acid rain, the dominant world-wide policy response has been one of inaction. Part of the explanation rests no doubt in the trans-jurisdictional aspects

[1] One such effort is under way at the National Center for Atmospheric Research in Boulder, Colorado.

of acid rain; a country (or city, or state, or province) has little incentive to mandate emission controls for the benefit of its neighbors. International institutions are currently too weak to promulgate regulations and secure compliance. An attempt has been made to specify uniform sulfur dioxide standards for the nations of the European Economic Community, but the unanimous assent of EEC members needed to bring regulations into force has yet to be achieved.

If attempts at transnational policy action have fared badly, policy responses for dealing with purely domestic aspects of the acid rain problem have been no more evident. Countries where government-sponsored studies have identified the need for pollution control to counter acid rain damage – such as West Germany and the USA – have done little more than countries such as the UK and France, which deny the existence of the problem. Laws that *are* on the books often go unenforced; a scrubber may be turned off with the flick of a switch. In the USA some of the worst offenders – copper smelters in Arizona – have been granted "extraordinary relief" from compliance with the federal Clean Air Act.

Far from solving the acid rain problem, existing public policies have if anything made matters worse. In some cases governments have required the installation of electrostatic precipitators to remove ash and suspended particulates from smokestack emissions. In other cases public policy has mandated construction of tall smokestacks in order to protect the local environment – only to ensure that pollutants spend a longer time in the upper atmosphere, thus promoting the likelihood of their oxidation into sulfuric and nitric acid. For example, the world's largest single source of acid rain is a nickel smelter in Sudbury, Ontario; where once the smelter had a number of small chimneys, today it sends all its emissions into the upper atmosphere via one huge smokestack.

Why did the acid rain problem arise? Why did it take so long for the problem to be recognized? Why is there such variability in and uncertainty over the causes and consequences of the phenomenon? Why can nobody determine whether acid rain is getting better or worse? Why has so little action been taken anywhere in the world to combat the problem? Why has public policy made matters worse rather than better? Why do informed individuals disagree so much about what (if anything) should be done? How might we think about effectively solving the problems related to acid rain?

This is not a book about acid rain.[2] But these questions, as they arise

---

[2] Readers interested in acid rain may find Postel (1985), Friedman and McMahon (1984), Wetstone and Rosencranz (1983), and Gorham (1982) instructive.

for acid rain, are typical of a wider range of ecological problems. Where, then, might answers be located?

Diagnoses and cures for ecological problems have, in fact, been sought in many locations. Different thinkers have blamed technology, indus- trialization, religion, social values, dominant worldviews or paradigms, and ill-chosen public policy. One finds corresponding variety in paths to salvation. The more widely charted (if less well trodden) include "appro- priate" technology, deindustrialization, religious conversion, a universal "paradigm shift" in the way we see the world, professionalized policy- making, moral re-education, more and better science and technology, and increased funding for environmental research.

This volume rests on the contention that answers to the questions enumerated above – for acid rain and for the more general class of ecological concerns – may be located in the mechanisms that contem- porary societies employ to make choices and solve collective problems. Thus the kind of world we create, both for dealing with ecological problems and for ordering society's affairs more generally, depends crucially on the structure of these mechanisms. My purpose is to locate deficiencies in the major existing mechanisms, and to see how we might do better. What exactly does this focus on collective choice entail, and how may it be justified?

SOCIAL CHOICE

A social choice mechanism is a means through which a society – whether local, national, supra-national, or global – determines collective out- comes (i.e., outcomes which can apply to all its members) in a given domain. The structure of social choice becomes especially interesting under conditions of non-uniform preferences among the members of that society. Indeed, some theorists define social choice in terms of procedures for aggregating conflicting preferences. This conceptualization delimits the field of axiomatic social choice theory (see, e.g., Arrow, 1963). It should be stressed that the question of social choice is interpreted far more broadly for the purposes of this study. Here, social choice will also concern the solution of collective problems. Whereas axiomatic social choice theory is confined to interpreting society's welfare in terms of some aggregation of individual preferences, my broader treatment allows for collective interests beyond these preferences (for example, an interest in ecological integrity).

The concept of social choice is not coterminous with governmental authority systems: governments constitute just one category – or, in some cases, just one component – of social choice. So, for example, many of the allocative outcomes produced by governmental structures could alternatively be determined by markets; and vice versa. It should also be noted that social choice mechanisms can have informal as well as formal components. Thus, social choice in most political systems proceeds in the context of constitutional rules; but in addition, all such systems possess informal channels of influence and communication, without which they could hardly operate.

Categories of social choice include hierarchies of command, markets, bargaining, anarchy, voting, legal systems, persuasion, and even armed conflict. Any real-world collective choice mechanism may contain a mix of elements from these categories. Real-world examples include the governmental system of the Soviet Union; the legal arrangements for environmental protection established under the US National Environmental Policy Act; the British electoral system; deterrence between the nuclear powers; the organization of a commune; and World War II.[3]

Social choice mechanisms can be centralized and cybernetic (i.e., consciously regulatory), centralized and purposive (i.e., actively and consciously pursuing goals), or decentralized and self-regulatory. Essentially, then, social choice mechanisms are society's collective means for coping with or solving problems.

Why focus on social choice? Some would argue that collective choice structures are merely derivative from technical, economic, or cultural arrangements. So, for example, Marxists see social and governmental forms as mere "superstructure," determined by the prevailing mode of production. Contemporary political scientists sometimes view governmental forms as the product of the "political culture" of a society. One of the more powerful analyses in the area of ecological concerns, that of Lovins (1977), sees any choice of what he calls a "hard" energy path as eventually leading to authoritarian and repressive centralized social, economic, and political structures, on the grounds that centralized energy is vulnerable and insecure, demanding strong central control to protect it.

---

[3] The term "regime" has recently gained wide currency in the literature on international politics. According to Young (1982), a regime has substantive components (rights, rules, regulations, and incentive systems) plus procedural components (social choice and compliance mechanisms). Social choice mechanisms in Young's sense cover only procedures for aggregating conflicting preferences; I prefer to use the term more broadly, and hence to treat regimes as just one kind of social choice mechanism. Regimes, as Strange (1983) points out, consist of institutional arrangements centered upon international bureaucracies and multilateral bargaining. The focus of the present study is of course much broader, even when it addresses the international system (chapter 12). Hence Strange's pointed critique of regime theory does not apply; indeed, this study is consistent with Strange's suggested political–economic alternative to regime analysis.

Conversely, Lovins believes a "soft" path will promote decentralized and market-like forms of social choice. Such technological determinism also informs the thinking of those who consider that "small is beautiful." It is worth noting here that, historically, small-scale technologies have formed the material basis of highly centralized empires (Bookchin, 1982: 240–1).

My contention is just the reverse: that the nature of the collective choice mechanisms in place will largely determine the kind of world that ensues. Certainly, that world is not *fully* determined by the mechanisms it contains. Yet, over the past three decades, it is clear that our collective choice structures have led us into situations which are in nobody's real interest and which very few people wanted, the threat of nuclear holocaust being the most obvious example. Public policy outcomes often bear only a remote connection to the intentions of "policy-makers" – even presidents sometimes feel themselves to be at the mercy of the forces surrounding them. Supposedly radical governments in the UK, France, and elsewhere rapidly run up against the "reality" of the economic system they must manage and the institutional structures in place to manage it; eventually, their practices come to differ but little from their more conservative predecessors. The market, still ubiquitous, if not the dominant form of social choice in the Western world, is indeed a kind of "prison" constraining the actions of governments. As Lindblom (1982: 326) puts it, "punishment does not depend on conspiracy or intention to punish": rather, the punishment, through economic downturn and unemployment, of any governmental actions which attempt to constrain or redirect a market system, is inherent in the very logic of market social choice.

Digging a little deeper, it is evident that societies characterized by relatively extensive market mechanisms produce a very different mix of goods for their members than societies with a heavier degree of more democratic, governmental social choice. Therefore, at comparable levels of national income per head, Americans get large quantities of private goods together with considerable public squalor (Galbraith, 1958), whereas Swedes get fewer private goods but far more in the way of public goods. Turning to the structure of political systems – a central aspect of social choice – the responses of different countries to the energy crises of the 1970s are instructive. In the fragmented American polity, tangible and immediate actions were directed at specific groups of sufferers; in the more centralized British system, an immediate increase in taxation on oil products was imposed, and direct controls were flirted with. More dynamically, institutional innovations can have a clear effect on the substantive content of subsequent choices: one need only look at the effects of the US Constitution. On a somewhat less grand scale, the policy process changes resulting from the passage of the US National Environmental Policy Act (NEPA) in 1969 have restrained the actions of development-minded federal agencies and forced them to contemplate the environmental aspects of their proposals and decisions. The NEPA framework conditions policy outcomes without anyone presiding over it, and resists attempts by actors such as presidents and secretaries of the interior to downgrade environmental concerns.[4] Social choice mechanisms do, in fact, generate (and are also, to a degree, the product of) convergent sets of expectations on the part of the actors involved in them, and so take on a life and momentum of their own.

AN ECOLOGICAL PERSPECTIVE

Social choice mechanisms are, then, crucial in determining the broad kind of future that unfolds. But to what degree may one justify dwelling on the specifically *ecological* aspects of this future? And, if one decides to take an ecological perspective, how might one go about assessing social choice mechanisms in such terms? Chapters 2, 3, and 4 take up these questions.

Beginning with the issue of justification, are ecological problems really so severe? Clearly, no society or civilization can survive for very long, let alone flourish, without a sound ecological base to provide necessary resources and crops, assimilate and recycle wastes, and buffer against extreme fluctuations in air and water cycles. That much about ecology is uncontroversial. The more interesting question concerns whether or not the capacity of local, regional, and global ecosystems to provide the basic needs of human (and non-human) existence is threatened. If no such stress is present or imminent, then an ecological inquiry into social choice is hardly worth pursuing.

The issue of severity is picked up in chapter 2, where I argue that ecological problems are indeed among the most severe challenges that the world faces. Cognizant of the arguments of those who deny this severity by pointing to improving trends on particular indicators, chapter 2 argues that such improvements are largely illusory. They result from problem *displacement* as much as from problem *resolution*. Displacement can proceed across space (for example, when a community exports its pollution), across time (as with long-lived toxic and nuclear wastes), and across media (for example, when air pollution is dealt with by

---

[4] I do not mean to suggest that the NEPA framework is perfect; it does, in fact, have a number of severe defects (see Bardach and Pugliaresi, 1977).

converting it into water pollution). As long as there are unlimited destinations for displaced problems, there is little cause for alarm. Unfortunately, the number of destinations proves finite.

The profound severity of ecological problems justifies a determination of whether or not each of the world's existing social choice mechanisms can cope with them. If we were lucky, we might be able to remedy any deficiencies in established mechanisms by a slight adjustment in the mix of different kinds of social choice. The most obvious such minimal adjustment would involve designation of an "ecological" issue area and the creation of effective mechanisms to deal with problems in this area. Dominant forms of social choice would therefore be left largely intact in other domains and policy issue areas. As will be shown in part III, there is perhaps some limited scope for piecemeal introduction of ecologically benign social choice mechanisms along such lines. Unfortunately, as shall be demonstrated in part II, existing dominant forms of social choice (such as markets and bureaucracies) make uneasy bedfellows by exporting their imperatives to other mechanisms – and with a vengeance to incipient forms trying to assert themseves.

Thus, if we take ecology seriously, and if the major existing mechanisms prove ecologically inadequate, then we should countenance the wholesale reconstruction of social choice mechanisms. So it is not just that the area of ecological issues demands a particular kind of social choice by virtue of the peculiar nature of the problems in this area: rather, these problems are both different and fundamental, and hence may lay claim to the basic structure of social choice. The priority of ecological concerns is addressed at length in chapter 5, which makes an ethical argument for according overarching importance to ecology.

But just how different are ecological problems? And what qualities do they demand of social choice mechanisms? These questions are the subjects of chapters 3 and 4. Chapter 3 begins by examining the general characteristics of ecological problems: complexity, non-reducibility, temporal and spatial variability, uncertainty, a collective quality (manifested in "public good" and "common property" problems), and spontaneity (the partial capacity of ecosystems to cope with stresses without human intervention). The idea of "ecological rationality" is then introduced as the capacity of human and natural systems in combination to cope with problems of this sort.

The implications of ecological rationality for social choice are explored in chapter 4. Five criteria for the assessment of social choice mechanisms are developed: negative feedback (the production of responses to human-induced shortfalls in life-support capability), coordination (across both actors and decisions), robustness (of performance across different circumstances), flexibility (in adjusting structure to cope with novel conditions), and resilience (the ability to correct severe disequilibrium). These criteria set the scene for the appraisals of particular mechanisms in parts II and III.

EVALUATION AND INNOVATION

The alternative forms of social choice which will be addressed in part II are enumerated in chapter 6. The classification developed is based on the distinct kind of *coordination* across actors which a mechanism embodies. (Coordination puts the "social" in social choice.) The resulting taxonomy consists of seven mechanisms which together dominate the contemporary world – and the intellectual discourse within this world: the market, administered systems, polyarchy, law, moral persuasion, bargaining, and armed conflict. Each of the first five has a chapter devoted to it in part II (chapters 7–11). Bargaining and armed conflict are most prominent in the international system, and so both are investigated in chapter 12 under the international rubric.

Despite their seeming disparity, not to mention appearance on opposite sides in contemporary debates between advocates of markets and governmental systems, these established mechanisms turn out to be remarkably similar from an ecological standpoint. All prove to perform poorly on the five criteria of ecological rationality. More important, these poor showings share a common root. All these mechanisms fall victim (if to varying degrees) to the tendency to displace rather than resolve ecological problems. *This tendency is inherent in their very structure.* Therefore any exhortation to society to change its ways (for example, by adopting better public policies) is likely to be frustrated unless the parameters of social choice are altered.[5] Moreover, any mix of these mechanisms is likely to provide only more extensive opportunities for problem displacement.

[5] Examples of such exhortation in the ecological realm are numerous. A plethora of works on energy policy evaluates and recommends courses of action for national governments. So, for example, Lovins (1977) suggests that governments promote and pursue a "soft energy path" composed of short-term and careful use of fossil fuels and long-term development of renewable, appropriate-technology energy resources; and that they simultaneously suppress a high-technology, centralized "hard path." In the area of natural resource management, Page (1977) recommends that governments adopt a set of policy instruments – notably, severance taxes – with the purpose of stabilizing an index of the real price of natural resources. This strategy is, for Page, the essence of conservation. In environmental policy, Commoner (1972) exhorts governments to undertake a program of wholesale ecological reconstruction. Ehrlich (1968) and Brown (1974) urge governments to pursue population control programs.

This universal failure justifies the explorations in institutional redesign pursued in part III. Chapter 14 delves further into the origin of problem displacement in contemporary social choice, and locates it in the kind of *reason* these mechanisms embody. Albeit in different ways, they all manifest instrumental–analytic rationality: the idea that problems are best resolved by an initial disaggregation into their component parts, each of which should then be attacked by formulating and effecting actions in pursuit of given ends. The idea of an "open society" is described in chapter 14 as the epitome of the *best* in instrumental–analytic problem-solving that contemporary kinds of social choice can offer. As a paragon that ultimately fails when seen in an ecological light, the open society acts as a foil for the agenda of institutional innovation developed in the remainder of part III.

The most prominent item on this agenda concerns the institutionaliz-ation of "practical reason" in social choice. Practical reason involves the rational scrutiny and generation of purposes as well as means, and proceeds pedagogically and communally rather than instrumentally and privately. It rules out the instrumental manipulation of social conditions and the pursuit of private ends in the public arena, accepting only a participatory, discursive kind of collective problem-solving.

Part III also explores the ecological attractions of radical decentraliz-ation in social, economic, and political organization. The prospects for both practical reason and radical decentralization are illuminated by reference to actual and potential experiments in innovative kinds of social choice. There proves to be more to such ideas than arid philosophi-cal speculation.

Social choice innovation in pursuit of ecological rationality and practical reason faces substantial obstacles in the world whose dominant imperatives are dictated by economic and political rationality. The latter are unlikely to provide much room for widespread innovation in pursuit of ecological rationality. The final chapter seeks encouragement from historical cases of redesign in social choice. Perhaps the most pressing problem in social choice today concerns the establishment of the preconditions for institutional redesign.

Let me now return to the beginning of this story, and consider the severity of ecological problems.

2

Is There an Ecological Crisis?

CRISIS? WHAT CRISIS?

One view of contemporary ecological problems has an "environmental crisis" occurring in 1970 and an "energy crisis" in 1973 (with an echo in 1979). The former was solved by prompt governmental action (at least in the Western democracies), the latter by market forces responding to increases in the price of oil. As I write, the prevailing mood is reflected less in complacency – it is still politic for government leaders to articulate a commitment to "cleaning up the environment" – than in a belief in the manageability of ecological problems. The result is that such concerns can be safely consigned to one or more functional areas, called environmental policy, natural resource management, or energy policy. Such areas, it is thought, are freighted with no more intractability and no more intrinsic significance than familiar categories of political concern such as health, defense, education, trade, transportation, housing, and so forth. Malthus is once again out of fashion, while at least some parts of the industrial world luxuriate in a renewed spurt of economic growth after the hiatus of the 1970s. These parts of the world tend to set the intellectual agenda for the whole (at least in social science). The pessimism of latter-day Malthusians has been swept aside by a renewed cornucopian vision of limitless possibilities.

This cornucopian faith is reinforced by a number of recent empirical accounts of man's increasingly harmonious relationship with the natural world, some polemical (J. Simon, 1981; Simon and Kahn, 1984), some more subdued in tone (e.g., Smith, 1979). Cornucopians can point to the following kinds of evidence to back their case that conditions have been improving – and will continue to improve:

(a) declining trends over the long term in the real prices of specific natural resources such as copper, tin, coal, and oil. "Real price"

here means price relative to other goods and services. Economics
tells us that price is a measure of scarcity; hence resources are
becoming less scarce;

(b)   worldwide increases in life expectancy: Julian Simon (1981: 6)
claims that life expectancy is the best indicator of total pollution
levels, hence pollution is decreasing worldwide;

(c)   increasing agricultural yields;

(d)   increasing global food supply in relation to the world population;

(e)   declining trends in indicators of specific pollutants (such as biological oxygen demand in British rivers, suspended particulates in
the air of US cities).

Malthusians, in contrast, see only worsening trends. They can counter-
attack with evidence such as the following:

(a)   declining proven quantities of specific resources in relation to rate of
resource use;

(b)   deforestation in relation to the remaining area of forest cover;

(c)   rate of topsoil loss in comparison to the regenerative capacity of the
land;

(d)   continuing buildup of the proportion of carbon dioxide in the
atmosphere (threatening climatological change through the "greenhouse effect");

(e)   drought in much of Africa (though the relative influence of natural
and human factors has yet to be ascertained; see Brown et al., 1985:
14);

(f)   the increasing quantities of fertilizers and pesticides necessary to
maintain crop yields;

(g)   the rate of extinction of plant and animal species, and concomitant
loss of genetic and species diversity;

(h)   human population growth rates;

(i)   specific environmental catastrophes, such as the 1984 Bhopal
disaster in India (in which at least 2,000 people died following the
accidental release of a cloud of methyl isocyanate gas), and the 1986
nuclear eruption at Chernobyl in the Soviet Union;

(j)   the increasing damage caused by acid rain to forests and lakes.

The debate between cornucopians and Malthusians might exasperate
any neutral observer, for there are so many empirical points at which
battle can be joined that clear victory for either side is unlikely. Total
victory would require one side or the other to carry its case on *every*
point at issue. Even then, one could not be sure that all the relevant issues
were covered, as new ones can arise; for example, awareness of acid rain
is of relatively recent vintage.

Unfortunately, *the empirical evidence cannot conceivably be decisive.* Mere examination of data does not allow one to judge the current severity of ecological problems, or the direction (let alone the slope) of trends in that severity. This impossibility will hold despite the availability of ever more and better data.

Empirical evidence is impotent here because an improving trend on a single indicator may result from nothing more than the export of difficulties to a problem area for which a different indicator is appropriate. Conversely, apparent deterioration on an indicator may simply reflect an extra burden imposed upon it from outside. This idea of problem displacement merits examination in depth.

PROBLEM SOLUTION OR PROBLEM DISPLACEMENT?

To illustrate how problem displacement can give the illusion of problem-solving, consider the following example, adduced by one of the most skilled polemicists on the cornucopian side of the ecological debate. Julian Simon claims that air pollution in the USA is decreasing with time (Simon and Kahn, 1984: 9). The best piece of evidence he can point to in support of this claim is reproduced in figure 2.1. The declining trend in sulfur dioxide pollution levels is especially noteworthy. But look at the small print: this declining trend is based on samples from urban sites only. Perhaps the urban sulfur dioxide problem *is* being solved – but how? Most sulfur dioxide comes from stationary sources (i.e., smokestacks). The problem has generally been "solved" by the construction of tall smokestacks. Thus, instead of polluting areas adjacent to copper smelters in Utah or coal-burning power stations in Ohio, the sulfur dioxide ends up in the form of acid rain in rural areas such as the Rocky Mountains or the Adirondacks. The problem has been exported or displaced rather than solved; it becomes "somebody else's problem."

In this example, displacement proceeds across space. But it can also occur from one medium to another and across time.

Spatial Displacement

Displacement across space may be illustrated further by the following examples.

(1) Leaking US toxic waste dumps are "cleaned up" with money appropriated to the federal "Superfund" in 1980. The hazardous wastes are removed to other dumps – supposedly more tightly sealed against extrusion into surface water or groundwater. Within a few years, some of

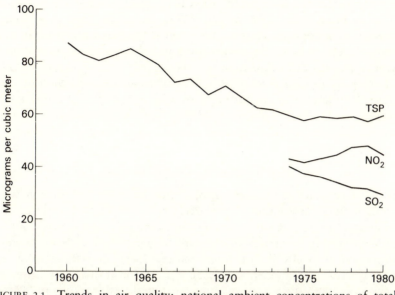

FIGURE 2.1   Trends in air quality: national ambient concentrations of total suspended particulates, nitrogen dioxide, and sulfur dioxide, 1960–80 (from Simon and Kahn, 1984: 11).

Data may not be strictly comparable. TSP (total suspended particulates data) for 1960–1 are based on 95 sites; data for 1972–6 are based on more than 3,000 sites; for 1977–80, there were 1,925 sites. The annual standard for TSP is 75 micrograms per cubic meter. SO_2 (sulfur dioxide) data are based on 84 sites, all in urban areas. The annual standard for SO_2 is 80 micrograms per cubic meter. NO_2 (nitrogen dioxide) data are based on 338 sites. The annual standard for NO2 is 100 micrograms per cubic meter. (*Source*: US Environmental Protection Agency)

the destination dumps have themselves started to leak. The further export of hazardous wastes to offshore incineration sites is under study.

(2)   Polluting industry has in some cases found it financially advantageous to relocate its activities in countries with lax environmental standards. Third World countries are often the "beneficiaries."

(3)   Southern California has insufficient locally available supplies of fresh water to support its growing urban areas and agricultural activities. So the area imports water via a system of aqueducts. In solving its water supply problems, southern California is creating ecological problems in the areas from which water is taken. One well publicized case is the Mono Lake watershed on the eastern side of the Sierra Nevada: Mono Lake itself is now on the verge of becoming an alkaline biological desert. Southern California also manages to avert the pollution and hazards associated with power production by ensuring that the coal and nuclear plants supplying its electricity are located in the desert to the east.

(4) Farmers in the world's tropical rain forest biome clear trees in order to plant crops. After a year of two, the cleared land is no longer capable of supporting crops. The farmer then moves on to unclaimed virgin forest, to begin the "slash-and-burn" cycle over again. Within a year or two it will be time to move on once more. The declining fertility of one piece of land is resolved by moving to another. A swath of "red desert" (incapable of supporting plant life) may be left in the farmer's wake.

Displacement to Another Medium

Displacement across physical media often arises in conjunction with attempts to treat pollution. In the acid rain case discussed above, the "solution" of electrostatic precipitation to the problem of smoke increases the acidity of emissions. A more frequent kind of cross-medium displacement backs the ecological adage that "everything must go somewhere" (Commoner, 1972: 36–7). An illustrative case arises in connection with the running example of Jones's (1975) study of "Clean Air." Jones chronicles the pollution problems associated with the huge US Steel coking plant at Clairton, outside Pittsburgh. The manufacture of coke yields a large amount of "process water" as a byproduct. This process water contains substantial quantities of toxic chemicals, leading to a potential water pollution problem. US Steel "solved" this problem by using the water to "quench" coke – the final stage in coke production, necessary to prevent further oxidation. Unfortunately, spraying process water onto hot coke caused an air pollution problem (which is Jones's (1975) concern). Eventually, US Steel reached an accord with Allegheny County which, for a fee, agreed to let the company discharge process water into the county sewers. The toxic chemicals are extracted from sewage by the county, and now accumulate in the form of sludge, for which no good disposal method has yet been found (see Mosher, 1983). Thus, what began life as a water pollution problem was converted into an air pollution problem and eventually into a toxic sludge problem; in no sense has it been "solved."

Cross-medium displacement occurs with resource exploitation too. Once a resource becomes scarce, its price rises (as long as there is any semblance of a private enterprise system). This price increase spurs a search for substitutes for the resource in question. Thus, a scarcity of wood in medieval Europe led to the development of coal as a fuel; a scarcity of oil today leads to the development of nuclear energy. Most resource economists view this displacement of scarcity across resources as positively desirable. But such a judgment rests on the faith that there will always be another resource to move on to.

Displacement to the Future

Ecological problems can be exported across time as well as space. The clearest example here may be long-lived nuclear wastes, a problem we bequeath to our descendants for the sake of meeting our own energy and weapons requirements. However, one well-known biologist (E. O. Wilson) believes that the crime for which our descendants will be least likely to forgive us is the genetic depletion attendant upon species extinction. Species extinction denies to the future the genetic material which could have uses such as improving the productivity and resistance to disease of crops, or providing new kinds of drugs.

Temporal displacement in connection with radioactive wastes and genetic depletion is a long-term phenomenon. More short-term manifestations of this kind of displacement may be found in the use of pesticides to increase crop yields. Pests can rapidly develop a resistance to particular chemicals, which means that the farmer must apply ever-increasing quantities of the pesticide (or ever more powerful versions of it) to maintain the same level of effect. Thus, today's "solution" is just the first step on a treadmill of recurring efforts to outpace the adaptive capacity of pest species.

Ecological problem displacement can occur across more than one dimension simultaneously. Consider again the acid rain case, where spatial displacement via tall smokestacks coexists with cross-medium displacement arising from the use of electrostatic precipitators to reduce the suspended particulates (ash) in emissions.

If my contention is correct — that we tend to export or displace ecological problems, rather than truly solve them — then clearly no indicator — *or finite set of indicators* — can tell us whether matters are really improving or deteriorating. Ideally, one would be led at this juncture to seek information on the marginal effect of each action on some single index representing "total global ecological integrity." But there are good reasons to doubt that the ecological theory needed to back the computation of such an index and the marginal impact of particular actions upon it could ever be constructed (see chapter 3). In the absence of any such index, it is hard to pass summary judgment on the tendencies toward problem displacement that I have identified. If ecological problems are unimportant, this tendency is nothing to worry about. If, on the other hand, they are life-threatening, then we should indeed be concerned. Yet the conundrum is that the available empirical evidence cannot tell us one way or the other. So just how severe are contemporary ecological problems?

Let me approach an answer by way of metaphor. Imagine an inflated balloon as representing the sum of our ecological problems. Any solution to a particular one of these problems may be likened to sticking one's finger into the balloon. If we get displacement rather than solution, the balloon will expand by an equal volume at some other point(s). We may press in a finger at this second location – only to find expansion at a third place; and so *ad infinitum*.

A cornucopian might respond by saying there is nothing to worry about here, for *we live in an infinite world*. That is, there will always be somewhere else (in the language of ecology, another niche) for the balloon to expand into. Thus, J. Simon (1981: 42) answers his rhetorical question, "Can the supply of natural resources really be infinite?" with a resounding "yes!" This position reflects a metaphysical commitment to the existence of ever more resources and "sinks" for pollutants as we use up current ones. In this spirit, Simon and Kahn (1984: 25) predict increasing energy availability "at least until our sun ceases to shine in perhaps 7 billion years." According to the cornucopian view, there will always be new "frontiers" to expand into.

On the other hand, what if the world is not infinite, so that there *are* limits somewhere? In this case, problem displacement will give the illusion of problem solution – until the limits are approached. This prospect is ominous, for ultimately nothing fails like the appearance of success. So: are there limits?

ARE THERE ECOLOGICAL LIMITS?

This question of limits is sometimes discussed in terms of "scarcity" – or, more broadly, "ecological scarcity" (for example, Ophuls, 1977). If one thinks in terms of the resources that human beings utilize – minerals, plants, animals, atmosphere, or assimilative capacity of the environment – then it might seem that ecological problems inhere in these things running out, or becoming increasingly scarce. So a number of neo-Malthusians foresee imminent collapse as the world and its economy encounter absolute physical constraints (see, for example, Meadows et al., 1972).

The term "scarcity" as generally used in economics refers to a condition under which available resources cannot satisfy all conceivable wants. As such, scarcity is the universal condition of humankind. But this form of scarcity refers only to individual wants in relation to specific resources, and not to society's total needs in relation to the absolute sum of resources.

"Particular" scarcities manifested in the depletion and exhaustion of specific resources are no cause for concern. If a specific resource "runs out," then a rational social order will simply develop a substitute (Hess, 1979: 73; Bookchin, 1982: 260–2; J. Simon, 1981). In other words, the scarcity will be displaced to another resource. Historically, no resource has truly "run out" for the world as a whole without a substitute being developed for it. To use the ecological terminology, no resource has ever proven to be a true "limiting factor" on human existence in its entirety. Our very presence on the earth today is proof. The evidence from particular societies and civilizations in the past is less encouraging. Clearly, societies *have* perished as a result of exhausting their ecological base. One need only look at the deserts which now cover the once-fertile "cradle of civilization" in the Middle East.

The only absolute scarcity in the universe – or subsets of the universe – is low entropy, or order. All closed systems are subject to the second law of thermodynamics: in the absence of any external input of energy, the system will deteriorate over time into "sameness." That is, the low entropy of the system is lost. This process is as simple as the mixing of hot and cold water to produce a tepid fluid, and as ineluctable ʌʌ the weathering of a building.[1] The destination of all entropic processes is the "chaos" of random distributions of matter and energy. Technically, "available" energy (capable of doing work) is converted into useless, "unavailable" energy.

Low entropy exists in many locations: in mineral structures, in physiological systems, in ecosystems, and perhaps even in human problem-solving capabilities. But there is only a limited physical supply of low entropy – order – on this planet (Georgescu-Roegen, 1971). And low entropy has no substitute. Three avenues for escape from the apparently finite nature of this world are sometimes suggested.

First of all, there is what Georgescu-Roegen (1979) calls the "energetic" position, seeking salvation in an energy source which does not involve any depletion of non-renewable resources, and which is sufficiently cheap and plentiful to act as a substitute for low entropy of materials. The apparent low entropy created by our civilization (in the form of capital, social structures, institutions, knowledge, and so forth) has itself been made possible by extensive energy use – that is, by depletion of the low entropy contained in energy resources. A plentiful

---

[1] For another example, think of a mixture of iron, air, and water in a closed container: eventually the iron oxidizes and is dispersed.

new energy source would enable "more of the same" in terms of social and technical progess, but without the attendant depletion of irreplaceable resources. Such a source would also enable mankind to gather and concentrate extremely dispersed raw materials, and open up massive possibilities for recycling materials (which otherwise can have a prohibitive energy cost). The most popular candidate for this source is fusion. But, even assuming that such a source could be found – which is far from certain – it would, ultimately, provide no escape, as the amount of waste heat dissipated into the environment would ensure that the biosphere succumbs to an entropic "heat death" (Georgescu-Roegen, 1979).

A second possibility seizes on the fact that the earth is not a truly closed system: it receives continuous inputs of solar radiation. However, if sunlight *can* be concentrated sufficiently to reverse material entropy (for example, through orbiting solar collectors which beam microwaves to earth), then we end up in the heat death just noted; if sunlight *cannot* be so concentrated, it enables no escape from the workings of the second law.

The third possibility is escape in a more literal sense, through the colonization of space and the exploitation of extra-terrestrial resources (O'Neill, 1978). While a technical possibility for a small number of people, space colonization requires too much in the way of terrestrial resources to provide an escape from the limits to growth for the vast majority of the earth's population, let alone the rest of the biotic community (Ophuls, 1977: 122).

Questions of specific scarcities (for example, of oil, or copper, or forests), though providing grist to the mill of less scrupulous prophets of doom, will be set aside in the remainder of this study. Inescapably lurking in the background of all ecological problems, though, is the question of general scarcity – of low entropy. The implications of entropy for human decision will receive further treatment in several of the chapters that follow.

The severity of ecological problems can, then, be captured in terms of the extent to which low entropy is being depleted. To what extent is humankind currently engaging in irrevocable depletion of low entropy in its corner of the universe? There is no simple answer. Human population and economic growth increase entropy, which must somehow be "pumped out" of man–nature systems. Clearly, human encroachment on natural systems is occurring and expanding at unprecedented levels and rates. One need only look at deforestation, the buildup of carbon dioxide in the atmosphere, desertification, and current rates of energy use. On the other hand, ecosystems can, to a degree, cope with stresses upon them, for they contain homeostatic devices to react to external perturbations, and adaptive mechanisms to adjust to external secular trends. The existence of homeostatic and adaptive devices acts to dampen fluctuations, reduce departures from an equilibrium state, and hence conserve low entropy. If, though, external interventions are numerous and severe, and are heaped upon one another as displaced problems accumulate, these mechanisms may eventually break down in the face of the simultaneous, multiple, and drastic corrections demanded by the effects of human industrial and agricultural activity.

The extent to which stresses of this severity are currently being imposed upon ecosystems is not free from controversy. Clearly, some ecosystems have already been stressed to the point of destruction, and history provides many examples of local and regional ecological collapse. On the other hand, few signs of global collapse are yet apparent; though, should the ecosphere eventually show signs of failing to perform, it may well be too late by then to take corrective action. And, as suggested above, problem displacement may be giving the illusion of success, both locally and globally. The difficulty in reaching any summary judgment as to how far we are currently from any limits reflects the substantial human ignorance of the workings of the world's ecosystems, and concomitant uncertainty over conditions and consequences of actions (see chapter 3 for more detail). It is not easy to gauge how close we are to the edge. But ultimately, only populations and species that have a net positive effect on their own environment and the ecological community of which they are a part can be expected to survive and flourish (see Wilson, 1980); nature is unlikely to grant an exception to man.

With this recognition, I do not mean to place myself in the camp of those who see only doom in the human future. Catastrophe is not inevitable; there are no limits which cannot be anticipated and coped with, be it by acceptance or evasion, for low entropy can exist in human problem-solving structures too. The possibility of ecological disaster may best be thought of as what Lasswell (1965) calls a "developmental construct" – a projected and unwanted state of affairs which can be prevented through collective action based on its anticipation.[2] Developmental constructs should function as self-denying prophecies.

CONCLUSION

The basic ingredients of this study – ecology (in the form of ecological

---

[2] Lasswell himself addressed a somewhat different developmental construct: the militarized "garrison state."

problems) and social choice — have now been introduced. The remainder of the book will be devoted to exploring the interplay of these two forces. This task is fraught with difficulty, inasmuch as I have already established that empirical evidence is of limited utility in assessing whether a situation — and, by implication, the social choice mechanisms operating therein — is satisfactory or unsatisfactory, getting better or getting worse in ecological terms. How, then, may one best assess and compare the ecological merits of social choice mechanisms?

The answer, it seems to me, is that one should look first at the general characteristics of ecological problems, with a view to determining the logical demands that their resolution (or amelioration) makes upon social choice mechanisms. Chapters 3 and 4 will atempt to generate a set of ecological criteria for the appraisal of social choice mechanisms.

3

Ecological Rationality in Principle

This chapter and the next will determine how one may assess and compare social institutions in terms of their capacity to solve ecological problems. Assessment and comparison in their turn demand criteria of evaluation. Therefore, I will attempt to develop, articulate, and justify a set of criteria for appraising what I call the "ecological rationality" of social choice mechanisms.

Ecological rationality is a form of "functional rationality" (Dryzek, 1983b; Bartlett, 1986). To describe a human social structure as functionally rational means, first and foremost, that its organization is such as to consistently and effectively promote or produce some value (see Diesing, 1962: 3; also Mannheim, 1940). Functional rationality, therefore, constitutes a standard for evaluation and design. A rational firm produces profits; a rational economic system satisfies consumer demand; a rational legal system resolves disputes; a rational collective security system safeguards peace. In this context, "reason is order or negative entropy, and rational norms are principles of order" (Diesing, 1962: 236). Any form of functional rationality embodies both a value (or values) and a mode (or modes) of behavior appropriate to the attainment of that value (or values). The purpose of this chapter and the next is to determine the nature of the value(s) embodied in ecological rationality, the modes of behavior required by that standard, and the qualities demanded of any social structure or social choice mechanism laying claim to ecological rationality. A logical prerequisite to that determination is a definition of the ecological context in which social choice mechanisms must perform. The next section therefore will attempt to capture the essential features of ecological problems. What, if anything, is special about the demands that ecology makes upon social choice?

THE NATURE OF ECOLOGICAL PROBLEMS

Any "problem" is simply a discrepancy between some ideal and actual (or projected) conditions, which is amenable to amelioration, if not elimination. *Ecological* problems concern discrepancies between ideal and actual conditions stemming from interactions between human systems and natural systems. (Just what constitute "ideal" conditions in this context will be determined in the next section.) The character of the natural systems in question, and the way in which they interact with human systems, impose, as I will now seek to show, some very special demands upon human problem-solving as that activity is extended to the ecological realm. A great deal is special about ecology. In the remainder of this section I will outline some distinctive features of ecosystems as entities to be confronted by human problem-solving processes, and go on to draw some implications for the nature of ecological problems.

Ecosystems

First of all, ecosystems exhibit a high degree of *interpenetration*. At the global level, such interpenetration is by now widely recognized: questions of environmental degradation, resource exhaustion, and population growth combine in what is sometimes termed a "problematique" (another useful technical term in this context is "mess"). Interpenetration holds at more localized levels too, however; as Commoner (1972: 29) observes, "everything is connected to everything else." Ecosystems themselves are always open systems, to the extent that it is often hard to define the boundaries of specific ecosystems. At all levels below the global ecosystem – or "ecosphere" – every ecosystem exchanges (to greater or lesser degree) both materials and energy with other ecosystems.[1] Hermetically sealed ecosystems exist only in the laboratory, and even these require a continual input of energy. So an element of arbitrariness in the definition of ecosystem boundaries is inescapable.[2]

The extent of interpenetration within and among ecosystems leads to a second striking feature of those systems: the ubiquitous presence of *emergent* (or "non-reducible") *properties* (see, for example, Edson, Foin, and Knapp, 1981). An emergent property is any characteristic of a system which is not predictable from a knowledge of the elements of that

[1] The ecosphere exchanges only energy with its environment; it receives an input of solar radiation, and radiates both heat and light into space.
[2] Ecologists themselves are not particularly troubled by this arbitrariness, for one can always specify both an "input environment" and "output environment" for any ecosystem that one chooses to define (see Odum, 1983: 16–17).

system. So the properties of carbon dioxide cannot be predicted from a knowledge of the properties of carbon and oxygen; the dynamics of a tropical forest cannot be predicted on the basis of a knowledge of each of the species living within it; and the extent of the life-support capacity of the ecosphere could not be predicted on the basis of a knowledge of its component ecosystems.

One of the emergent properties of ecosystems – sufficiently important to warrant classification as a third striking feature – is their *self-regulating* quality. Self-regulating systems are sometimes termed "cybernetic." While some observers cast doubt on the cybernetic nature of ecosystems (for example, Engelberg and Boyarsky, 1979) on the grounds that ecosystems lack any central goal-setter or distinct communication network, ecosystems are cybernetic, if only by analogy rather than homology (Patten and Odum, 1981: 888). Ecosystems do possess stable "goals," such as production–respiration ratios, total biomass, and species diversity, and there is a (sometimes elusive) information network composed of the "invisible wires" of nature (Patten and Odum, 1981). Those "wires" are everything that regulates the flow of matter and energy through ecosystems – including cues of sight, touch, smell, and temperature, and laws such as those of gravity and the conservation of mass and energy (Patten and Odum, 1981: 890). Ecosystems are "non-teleological": no central controller sets goals, monitors feedback, and acts in response. Instead, control devices are internal and diffuse.

So, for example, if a species threatens to explode in population, a predator's numbers may increase in response; if essential nutrients become impoverished, microbial subsystems may react so as to release stored nutrients into the system; if ambient temperature increases enough to render one kind of essential bacteria inactive, another kind may multiply and swing into action in performing the same function; if the carbon dioxide content of the atmosphere increases, then so does the photosynthetic rate of green plants which can, in part, absorb that increase.

Self-regulation enables ecosystems to maintain their essential structures and functions in the face of exogenous shocks. This homeostatic quality means that ecosystems can, to an extent, remain intact (as far as their "goal variables" are concerned) when confronted with the intrusions of human agricultural and industrial activity. Thus, nitrate pollution from agricultural runoff may lead to increased algae growth in a lake; that growth will in its turn be consumed by herbivores, until the excess of algae is removed. Further changes will ramify throughout the ecosystem of the lake, but eventually the system can converge on its pre-pollution equilibrium.

Self-regulation means, too, that ecosystems can adapt to permanently changed conditions in their input or output environment. So, for example, if a pond receives a continuous input of nitrates from its surrounding fields or sulfates from acid rainfall, then the species it contains, their relative numbers, and the character of their interactions may shift to a new equilibrium sensitive to the rate of nitrate (or sulfate) input. One possible long-term reaction to changed conditions is an evolution of physiological characteristics and behavioral traits through natural selection.[3]

Aside from homeostasis and adaptiveness, the dynamic quality of ecosystems is manifested in the process of "succession," a spontaneous developmental process involving changes in species composition with time. Thus, "pioneer" ecosystems (generally containing but a few species and interactions) gradually give way to more complex forms, culminating eventually in "climax" ecosystems. During succession, the physical environment itself can be modified; for example, a sand dune can be converted into rich soil.

Human activity in the ecological realm proceeds, then, in the context of interpenetrated, dynamic systems which exhibit emergent properties. Dynamism is manifested in homeostasis, adaptiveness, and succession. Ecological problems themselves typically arise as a consequence of human interactions with these natural systems. The circumstances of human problem-solving in the sphere of ecological concerns may, given the above-noted characteristics of ecosystems, be captured as follows.

Ecological Problems

Perhaps the most striking feature of ecological problems is *complexity*. Complexity may be defined in terms of the number and variety of elements and interactions in the environment of a choice process (whether that process is a human brain, a computer program, a government agency, or a market). Following Weaver (1948), it is useful to distinguish between "organized" and "unorganized" complexity. Under unorganized complexity, large numbers of elements interact randomly and in a systematically unrelated way; there is no great variety in interactions. Hence the behavior of the aggregate (though not of any one component) can be predicted statistically. Take, for example, consumers and producers in a free market, insects in a swarm, molecules in a liquid, or seedlings in an agro-ecosystem. In contrast, organized complexity involves systematic and variable interactions among the

---

[3] Note, for example, how readily "pests" adapt and become immune to pesticides.

elements of a system. Examples include small human groups, multi-national bargaining systems, and oligopolistic markets. The behavior of such systems cannot readily be captured on a statistical basis, let alone subsumed under general laws.[4]

Organized complexity may be expected in any system with purposeful – "teleological" – components. Any such elements are capable of devising the actions they take and therefore some of the interactions they make. Clearly, many human systems are teleological. Ecosystems themselves are non-teleological in nature; they have no supervising intelligence (except when they are managed by human beings). However, ecosystems do contain teleological elements (organisms and populations). Moreover, at the system level, "goal" variables such as production–respiration ratios and species diversity can be interpreted as elements in an "objective function" (Patten and Odum, 1981: 888). And a further teleological aspect will be introduced inasmuch as interactions are mediated by human beings, or human social systems. Therefore one can expect organized complexity in the ecological circumstances of human choice, especially in the interactions of natural systems and human systems. It will be shown in some of the chapters that follow that the extreme degrees of complexity characterizing ecological problems are potentially devastating to some of our familiar conceptions of problem-solving.

A second important circumstance of human choice in the ecological sphere is that ecological problems will often be *non-reducible*. By a non-reducible problem, I mean one whose resolution or amelioration cannot be guaranteed through resolution of its parts. This circumstance follows directly from interpenetration and emergent properties in ecosystems.

Interpenetration explains much of the problem displacement described in chapter 2. Take, for example, the problem of "pollution" in a given geographical area. One might, reasonably enough, proceed to divide the problem into parts such as air pollution, water pollution, and solid wastes. The water pollution problem could be "solved" by collecting and incinerating all waterborne wastes – but the consequence is increased air pollution. Air pollution from stationary sources can be ameliorated by electrostatic precipitation – which generates solid wastes. Solid wastes can be disposed of by mixing them into a slurry and discharging them into a watercourse – causing water pollution. To take

---

[4] Modern social science often does attempt just such a subsumption; as, for example, in statistical accounts of the determinants of voting behavior. Such efforts frequently have a procrustean air about them.

another example, consider how shortages of a key mineral resource might be alleviated. One solution would be to make use of lower-grade ores – but this use will be at the expense of increases in both energy input and pollution.

Emergent properties in ecosystems lead directly to non-reducibility in ecological problems. So, for example, at the level of the ecosphere as a whole there exist global cycles in water, carbon, and nitrogen. The properties of those cycles cannot be understood solely in terms of a knowledge of the various terrestrial and oceanic ecosystems. Any problems concerning those cycles – for example, the current rate of increase in atmospheric carbon dioxide, threatening a "greenhouse effect" – inhere in the cycle in its entirety, rather than in any of its component ecosystems.

It should be stressed, though, that a recognition of problem non-reducibility does not demand a *metaphysical* commitment to holism. At a purely pragmatic level, it is not easy to infer the properties of a truly complex system from a knowledge of its parts. Attempts to model the totality of interactions between the members of even the simplest ecosystem have met with no success (see Richerson, 1977: 8). As Herbert Simon notes, an attitude of pragmatic holism can be quite consistent with a metaphysical commitment to reductionism (H. A. Simon, 1981: 195).

Non-reducibility in ecological problems can make the way we customarily allocate problems into separate issue areas look suspiciously like a bad case of tunnel vision. Such allocation is likely to lead to problem displacement or export, rather than problem resolution. But it must be stressed that non-reducibility does not imply that the resolution of ecological problems demands centralized, governmental action – or, indeed, any conscious coordination in problem-solving. That question is, for the moment, an open one, one that will be addressed in several of the chapters that follow.

A third striking feature of ecological problems as they confront human systems is their temporal and spatial *variability*. The dynamic tendency of ecosystems – manifested in homeostasis, adaptiveness, and succession – ensures that the ecological context of any problem will not remain fixed. Interpenetration adds another layer to temporal variability by allowing events elsewhere in the ecosphere – either human or natural forces – the potential to introduce exogenous shocks into the domain of any problem of interest. Such shocks may take the form of temporary disturbance – as, for example, from a volcanic eruption or nuclear explosion – or permanent alteration in environmental conditions, such as changes in atmospheric composition, and creeping desertification.

Temporal variability can also arise as a direct consequence of problem displacement; for example, if a water pollution problem is solved by converting it into an air pollution problem, the latter will appear novel. *Spatial* variability in ecological problems arises because ecosystems, their components, and human systems can combine in diverse ways in different environments. Thus, oil pollution in the Arctic Ocean is a very different matter from oil pollution in temperate waters; deforestation in Central America involves a different set of considerations than deforestation in South East Asia. The extreme local variety in how acid rain problems are manifested was noted in chapter 1.

A fourth characteristic of ecological problems is *uncertainty*. Uncertainty can concern the nature of present or future conditions, and the consequences of human actions. Severe uncertainty can prevent probability judgments from being assigned with any confidence.[5] At a still more profound level, one may even be uncertain as to the range of possible conditions and consequences. Uncertainty is a product of interpenetration, emergent properties, and dynamism – in isolation or in combination. The more complex a system, the less "knowable" it becomes; the more it exhibits emergent properties, the less readily is it decomposed into simpler parts; and the more dynamic the system, the harder it is to capture its present state or predict its future course. Again, the acid rain issue illustrates the degree of uncertainty inherent in ecological problems – knowledge of the causes and consequences of acid rain is highly incomplete.

A fifth prominent feature of ecological problems stems as much from the character of human systems as from that of natural systems. Ecological problems are often *collective;* that is, large numbers of actors have a stake in them. A collective action problem exists whenever rational individual actions fail to produce a rational whole for society. The two most familiar kinds of collective action problem arising in the ecological realm are the "tragedy of the commons" and the underprovision of public goods.

The "tragedy of the commons" is the popular term (coined by Hardin, 1968) for a situation that often obtains under a regime of common property. A system of common property exists when large numbers of individuals are allowed unrestricted access to a resource. The world is well stocked with common property resources: examples include the assimilative capacity of a polluted river; the productive resources of ocean fisheries, migratory wildlife, and groundwater basins; the protec-

---

[5] Technically, it is this lack of potential for probability assignment which distinguishes uncertainty from "risk."

tive value of natural environments; and, at the largest scale, the carrying capacity of the ecosphere as a whole. The essence of a common property resource lies in rights to use the resource divorced from the responsibility for maintaining its quality. The ensuing incentives are such that rational, self-interested actors will each attempt to maximize their "take" from the resource, feeling (quite correctly) that if they fail to do so then other actors will reap the reward of intensified use. Each actor receives the entire benefit of its own increased use, but shares the cost of that increase with all other actors. Further, no actor has any incentive to look to the overall quality of the resource; any such investment would produce benefits shared by all the users, with costs falling totally on the investor. The overall result is overuse and depletion of the resource. This result will apply even if all the actors involved are committed to ecological values and are aware of the degradation of the commons in question. The essence of commons problems is not ignorance: it is solely a matter of the structure of incentives, and that is why the result is a true "tragedy". Indeed, even if loss of livelihood of those depending on a resource threatens, the tragedy will still ensue.

As human population and economic growth proceed, the number of resources whose common property nature becomes apparent multiplies. So, for example, ozone in the stratosphere is a common property resource currently being depleted as a byproduct of high-altitude supersonic flight and the use of fluorocarbon chemicals. The atmosphere as a whole is a kind of commons: each individual burner of fossil fuels uses up a small portion of its capacity to buffer carbon dioxide content, thinking little of his contribution to the greenhouse effect and climatological change. Families producing large numbers of children as a source of labor and social insurance contribute to the burden on global food supplies. American farmers "mine" groundwater reservoirs, knowing full well that depletion spells long-term disaster for them. On a larger scale, it is meaningful to speak of the earth's resources in their totality – and the low entropy they contain – as a kind of "global commons."

Collective action problems also arise in the case of public goods, which must be supplied to large numbers of people simultaneously if they are to be supplied at all, and the benefits of which cannot accrue exclusively to those purchasing them. These two qualities – jointness of supply and non-excludability – define the category of public goods.[6] The standard example of a pure public good is national security. Given that individuals cannot be excluded from the benefits of a public good, any economically

---

[6] See Head (1974) for an elaboration of the concept of a public good and its consequences.

rational actor will try to take a "free ride" by relying on the other beneficiaries of the good to pay for it. This attitude is economically rational because any contribution that an individual gives (or withholds) will make virtually no difference to the amount of the good he or she receives. If they are economically rational, all potential beneficiaries of the good will think in the same terms. So supply of the public good is problematical.

Within the sphere of ecological concerns, public goods include:

the integrity of the ecosphere;
the global supply of low entropy;
the quality of the global atmosphere;
biotic diversity;
the survival of the human race;
the quality of the oceans;
the quality of local environments;
knowledge of ecologically benign technology;
social choice mechanisms for solving commons problems in markets.

Clearly, this is a substantial list.[7]

The five features of the ecological problems described above seem to bode ill for human problem-solving. A sixth characteristic in the ecological circumstances of human choice alleviates this dismal prospect somewhat. This feature may be termed *spontaneity:* the capacity of ecosystems to cope with problems without human intervention. Spontaneity is a direct consequence of homeostasis and adaptiveness.[8]

These, then, are the ecological circumstances of human choice: complexity, non-reducibility, variability, uncertainty, collectiveness, and spontaneity. Having established these circumstances, the next step in the development of a set of criteria for social choice is the specification of some normative judgment. In other words, *why* should we care about ecology?

THE IDEA OF ECOLOGICAL RATIONALITY

The earth's natural systems and their elements can be regarded as valuable for a wide variety of reasons. These systems yield mineral,

[7] In passing, it should be noted that many public goods are the obverse of common property resources: the "good" is the quality of the "resource."

[8] The presence of such mechanisms leads some ecologists to advocate a policy of "nature knows best" in the resolution of ecological problems, to the extent that human problem-solving should take a back seat, if not leave the vehicle altogether. This argument will be considered in chapter 4.

renewable, and genetic resources for human productive activity.[9] In addition, natural systems treat and recycle human waste products. Through gas exchange, nutrient recycling, water purification, and temperature control, the environment stabilizes ambient air and water quality and temperature. Parts of the environment have amenity value, inasmuch as they yield human enjoyment. (Organizations such as the Sierra Club in the USA and the National Trust in the UK promote environmental amenity values.) There is considerable scope for treating nature in aesthetic terms (Sagoff, 1974). Non-industrial societies often attach spiritual and religious significance to their environment. Colonists spreading "civilization" may have seen the environment in terms of a challenge to conquer. (Think, for example, of romanticized accounts of the history of the American West.) Industrial societies and their scientific communities have questions to which nature may provide some of the answers – or even some of the questions. Flora and fauna, species and ecosystems, can be treated as valuable by virtue of their very existence, and therefore as having moral rights against human encroachment (Stone, 1972). (Animal liberationists and organizations such as Greenpeace embody this perspective.) Man and nature may be conceived of as constituting in effect a single community, to which man has obligations and commitments by virtue of membership (Leopold, 1949). Nature provides the potential for cooperative co-evolution of human and natural systems (cf. Tribe, 1974). Ecology can also be regarded as a *source* of values such as diversity, homeostasis, and adaptiveness (Lemons, 1981), or of a thoroughly non-hierarchical view of the world (Bookchin, 1980, 1982).[10]

In this study I wish to stress only the productive, protective, and waste-assimilative value of ecosystems – that is, those aspects which provide the basic requirements for human life. Henceforth, ecological rationality will be interpreted in these terms. Productive needs include renewable resources, non-renewable resources, and agricultural products.[11] Protection refers to the stabilization of man's ambient environment through the buffering of air and water cycles, the moderation of temperature extremes, and the regulation of the abiotic environ-

[9] Many resources remain to be recognized as such, given that man has yet to find a use for them (see Martin, 1979).

[10] Ecologists speak of "hierarchies" in nature, but these are hierarchies in the sense of levels of organization – from the gene to the ecosphere. Hierarchy in the sense of domination of one population by another cannot exist in ecosystems. For a more extensive survey of reasons for valuing nature, see Ralston (1981).

[11] Technically, the primary productivity of an ecosystem is the rate at which it can convert radiant energy into biomass. Here, I wish to restrict the concept to production for human uses, and to include mineral resources too.

ment (for example, atmospheric ozone and physical substrata). Waste assimilation is simply the recycling of pollutants.[12]

By no means do I wish to suggest that reasons for valuing the environment over and above its human life-support capacity are unimportant. An anthropocentric life-support approach is taken here for two reasons. First, a life-support approach is a "minimal" one. Introducing other reasons for attaching positive value to natural systems can only make the arguments that follow apply *a fortiori*. Second, in restricting oneself to some basic human interests, one can meet competing forms of functional rationality (whether economic, social, legal, or political) on their own ground: the ground of specifically human interests. Ecological rationality can therefore be made commensurable with its opposition. In chapter 5 I will argue for the priority of ecological rationality over these competing forms, an argument that can be put only if one takes an anthropocentric, life-support approach.

In these terms, how may one recognize an ecologically rational structure? As noted above, a functionally rational structure is one that is highly ordered. Setting aside for the moment the question of human interest, an ecologically rational *natural* system is one whose low entropy is manifested in an ability to cope with stress or perturbation, so that such a structure can consistently and effectively provide itself with the good of life support. The rationality of an ecosystem is therefore closely related to the quality of self-regulation, introduced in the previous section. An ecosystem's "objective function" applies to the wellbeing of the system as a whole, rather than to any specific individuals, populations, or species within the system.

Self-regulation, as noted above, is manifested in both homeostasis and adaptiveness. Homeostatic stability, in turn, can take two forms: resistance and resilience. Resistance stability is steadiness in reaction to stress. Resilience stability is the capability to recover quickly after stress has succeeded in removing the system from its normal operating range. Both resistance and resilience stability require negative feedback – that is, input to a system of signals that counteract deviations in its output. In an ecosystem, negative feedback inheres in the diffuse and internal controls discussed in the previous section. To a certain extent, functional redundancy can substitute for negative feedback in securing resistance and, to a lesser degree, resilience.[13]

[12] The term "waste disposal" is inappropriate, for "everything must go somewhere." For greater detail on production, protection, and waste assimilation in the human environment, see Odum (1983: 491–9).

[13] For greater detail on resistance and resilience, see Odum (1983: 47–53).

So much for the rationality of ecosystems *qua* ecosystems. But where do human interests fit into this scheme of things? After all, the concern of this study as stated is with what ecosystems can do for humans, rather than with what ecosystems can do for themselves. Ecological rationality for the purposes of this investigation is the capability of ecosystems consistently and effectively to provide the good of *human* life support. "Consistently" here refers to long-term sustainability. In other words, the wellbeing of the present generation should take no necessary precedence over that of future generations. This does not require substantial prescience in our present choices: all we in the present need do is conserve low entropy in human and ecological systems, so that we may pass on to our successors as much "order" as we ourselves started with.

Clearly, the productive, protective, and waste-assimilative requirements of human existence make some special demands upon the homeostatic and adaptive capacities of ecosystems. The more negative consequences of these demands constitute what appear from the human perspective as ecological problems – that is, actual or potential shortfalls in the human life-support capacities of ecosystems.[14] Such shortfalls may result from resource extraction and environmental pollution. Ecological problems may, as I have already noted, be located at the intersection of human systems and ecosystems. From the perspective of ecological rationality as defined, then, what one is interested in is the capacity of human systems and natural systems *in combination* to cope with human-induced problems.[15]

An ability to cope with or resolve problems may be termed "intel-

---

[14] This capacity is sometimes referred to as "carrying capacity." Carrying capacity for ecologists refers, though, to the total amount of biomass an ecosystem can sustain, not to specifically human biomass.

[15] The problem-solving capacities of human systems are, of course, of considerable interest beyond the ecological realm. There exist threats to human survival unrelated, in the first instance, to the dynamics of ecological collapse. Foremost among these is, perhaps, the threat of nuclear holocaust. In response to such threats, societies need to develop capacities to steer a secure course into the future. Ecological rationality, if achieved, enhances the likelihood – but does not guarantee – that such a course will be charted.

The idea of "strategic rationality" is sketched by Werbos (1979) and Dryzek (1983d) as a guiding principle for societal self-direction in the face of potential catastrophe. Essentially, strategic rationality requires that all decisions in the present be subordinated to the goal of maximizing the strategic problem-solving position of the next time period (a year, a decade, or a generation). Keeping options open is crucial. Strategic rationality and ecological rationality are, I believe, quite compatible. Their difference lies in the fact that strategic rationality is a form of instrumental reason, intended as a guideline for the substantive content of self-conscious collective choices. Ecological rationality is a functional counterpart of strategic rationality.

ligence." John Dewey recognized that "our intelligence is bound up . . . with the community of life of which we are a part" (Dewey, 1922: 314). So intelligence can be an attribute of organization: it is "the ability of a whole society or community to effectively solve the problems confronting it" (Diesing, 1962: 234).[16] The community in which Dewey and Diesing are interested consists only of human beings; the community of interest in this study consists of human and natural systems in combination.

GENERALIZATION ACROSS ECOLOGICAL CONTEXTS

On what combination of human and natural systems, then, should one seek to pass judgment? At one extreme, interdependence and non-reducibility in ecosystems and ecological problems might lead one to conclude that meaningful analysis is possible only at the level of our civilization as it combines with the ecosphere. Some broad-scale works follow exactly this approach.[17] If disaggregation were deemed desirable, one could perhaps decompose on the basis of biomes, and analyze social systems in the context of boreal forest, temperate grassland, desert, chaparral, and so forth.[18] At a more disaggregated level still, one could define one's unit of analysis in terms of local ecosystems.

Afficionados of the case-study approach to questions social and ecological would probably wish at this juncture to turn their attention to ecosphere, biome, or ecosystem and to the totality of the human systems that each contains. There is a long tradition of case study in both biological and human ecology. Defenders of this tradition might claim that the peculiarities of each case are sufficient to preclude cross-case generalization, and that complexity and non-reducibility in ecological systems render each case truly idiosyncratic, such that case-specific knowledge is the most powerful kind. Proponents of the case-study approach would be uneasy with any abstraction of social choice mechanisms from the specific contexts in which they operate.

---

[16] Note that good outcomes in response to problems can be achieved in a functionally rational system without any actor pursuing them consciously or instrumentally. The example cited most frequently in this context is that of "invisible hand" market mechanisms.

[17] See, for example, Toffler (1981), Henderson (1981), Rifkin (1981), Harman (1976), Hawken, Ogilvy, and Schwartz (1982), Heilbroner (1980), Kahn et al. (1976), Johnson (1979), Sibley (1977), and the work of the World Order Models Project (Mendlovitz, 1975).

[18] For an attempt, in part, to identify the forms of social structure appropriate to Arctic tundra and boreal forest biomes, see Dryzek and Young (1985).

However, case studies in both biological and human ecology themselves involve classification, decomposition, and abstraction from context. In biological ecology case studies, spatial boundaries are drawn around the edges of the ecosystem under study. Thus the ecology of a meadow may be studied in isolation from that of an adjacent forest – even though interactions are likely to be especially rich at the "ecotone" where meadow and forest meet. (As Leopold (1933) notes, "wildlife is a phenomenon of edges".) The case studies of human ecologists draw boundaries around specific populations of people, downplaying the interactions that each population makes with environments or social groups outside its immediate locality. So, for example, a plethora of studies in ecological anthropology examines how specific populations develop customs and behavioral patterns to promote their adaptation to the local environment. These studies have been extensively criticized on the grounds that they ignore the broader interactions made by the population in question (see Orlove, 1980: 244). We live in a world of "open" ecological and social systems. Drawing boundaries around local populations is no more defensible than drawing boundaries around local ecosystems (see Ellen, 1982: 78–9).

One of the few routes to valid knowledge is through classification, disaggregation, and decomposition (even when one is trying to expose the hazards of such procedures in a larger setting, as I will in this study). Indeed, this is exactly what the term "analysis" means. The interesting question is less *whether* one disaggregates, than *how* one does it. While biological ecologists delimit their studies with geographical bounds around ecosystems, and human ecologists with spatial bounds around populations, I prefer to delimit on an analytical rather than a spatial basis. That is, I will be extracting social choice mechanisms from their contexts to apprehend them as analytical types. This approach to analysis may be justified on the grounds that real-world social choice mechanisms straddle both ecological and cultural boundaries (boundaries which are themselves highly indistinct). For example, the administered system of the Soviet Union intersects very directly with tundra, taiga, temperate grassland, desert, and a variety of anthropogenic ecosystems, not to mention a number of very different social and cultural groups. Indirectly, that system interacts with a further range of ecosystems and societies outside the borders of the USSR. It is worth seeking information about the ecological problem-solving capability of that administered system in its totality, not just in terms of how it works out in Kamchatka or Kiev. The kind of knowledge sought here pertains to the ecological rationality of a social choice mechanism as it makes a variety of combinations with natural systems, social groups, and other social choice mechanisms.

I recognize that the kinds of generalizations I will be pursuing henceforth may not sit well with relativists who believe that all cases are fundamentally different. To assuage them, in the pages that follow I will make reference to numerous real-world combinations of ecosystems, ecological problems, and social choice mechanisms. However, the analysis will be case-sensitive, rather than case-specific. Relativists who remain unconvinced might still find parts II and III of this study useful as a prelude to case study; moreover, the ecological rationality standard as developed is applicable to spatially bounded cases, as well as to social choice mechanisms.

Any social choice mechanism is likely to make a variety of interactions with natural systems, social systems, and other forms of collective choice. One aspect of its ecological rationality must therefore be its ability to cope with variety in these interactions, and to form a rational whole in the various combinations it makes. Henceforth, I will speak in terms of the ecological rationality or otherwise of a social choice mechanism; it should be borne in mind that this usage is shorthand for "the ecological rationality of a social choice mechanism as it makes a variety of combinations with natural systems."

4

Ecological Rationality in Practice

FROM ECOSYSTEMS TO SOCIAL SYSTEMS

How, then, does one recognize an ecologically rational social system? One school of thought would argue that ecological and social systems are identical in enough important respects to render them amenable to analysis in the same terms (see, for example, Duncan, 1964; Michelson, 1970; Margalef, 1973). Attempts to unify social and biological ecology can draw epistemological support from the "general systems movement," whose intellectual goal is (or was) the unification of the study of all kinds of systems, whether social, physiological, ecological, or mechanical (see von Bertalanffy, 1968).

Parallels between ecosystems and social systems were explicit in the "Chicago School" of urban sociology (see, for example, Hawley, 1950), whose members analyzed cities and other social institutions in terms of ecological metaphors and analogies. For example, change in the demographic characteristics of an urban neighborhood could be described as "succession." Such parallels also arise – if less explicitly – in functionalist sociology (see Martindale, 1960). Functionalist sociologists explain social institutions and processes through reference to their contribution to the stability of society – just as ecological structures and relationships can be accounted for in terms of their contribution to homeostasis. This kind of "superorganismic" view of human societies and ecosystems is today out of fashion in both sociology and biological ecology. Its fall from favor stems in large measure from a questionable attribution of goal-directed behavior to social and ecological aggregates (see Richerson, 1977).

If the homology of social systems and ecosystems were taken to heart, the search for ecological rationality in human affairs would consist of an identification in social systems of ecological constructs such as homeostasis, succession, adaptiveness, and mutualism – the factors which maintain the "rationality" of biological ecosystems. This approach will not be followed here, for clearly, human social systems are not exactly like ecosystems. If nothing else, they contain only one species: man. It is not very helpful to think of human social groups as different "species" (cf. Ellen, 1982: 68). Social systems and ecosystems may have their parallels, but so do social systems and electrical systems.

A somewhat different linking of human and biological ecology does indeed take note of this central fact that man is a single species, and treats him like any other. Thus, sociobiologists interpret human social behavior in terms of genetic motivation; and their analyses of human society are therefore essentially identical to their analyses of insect society. In both cases, the essential motivation of individual behavior is assumed to be the maximization of genetic fitness (see Wilson, 1975, 1978; Pugh, 1977). Hirshleifer (1977: 2) points to a "striking parallel" between the "fundamental organizing concepts" of economics and sociobiology. He finds "isomorphism" between (human) optimizing and (non-human) adapting; between progress and evolution; and between innovation and mutation.[1] Sociobiologists apply biological concepts to human social systems, but moves in the reverse direction are also possible. Economic models are widely used in biological ecology, which is sometimes referred to as "nature's economy" (see, for example, Ghiselin, 1974). Some of the more recent applications of economic thinking to biological population dynamics involve the use of game theory to explain the evolution and selection of behavioral strategies (Axelrod and Hamilton, 1981; Maynard Smith, 1982).

Sociobiology has tended not to address the interactions that a species (be it human or otherwise) makes with its environment (though such study remains a theoretical possibility). Interactions of this kind have more often been explored by investigators who have located human populations in particular biological ecosystems, to be treated as just another component of the ecosystem in question. This approach to man in nature is found in the field of ecological anthropology, which strives to explain the cultural features of a given human society or population through reference to the natural systems with which it coexists (see Ellen, 1982; Orlove, 1980). A "functionalist" school within ecological anthropology regards *all* social and cultural characteristics as functional adaptations to the carrying capacity of the local environment (see, for example, Harris, 1966). From this viewpoint, *all* human societies are ecologically rational (and so a study such as this would be regarded as superfluous). However, less rigid versions of ecological anthropology

---

[1] Hirshleifer also, mistakenly, equates biological ecology and sociobiology.

allow ecological dysfunction in society and culture (for example, Butzer, 1980).

This linking of social systems and ecosystems by treating human populations like any other component of an ecosystem is controversial. Sahlins (1976) argues forcefully that culture in human society can have a life of its own, independent of ecological influences. Moreover, what Bernstein (1981: 325) calls "perceptual failure" is today endemic in human social systems; quite simply, people do not perceive the ecological consequences of their actions. (This failure arguably underlies many contemporary ecological problems.) Given the pace of social, economic, and cultural change in the contemporary world, it is entirely possible that social, economic, and cultural systems can evolve a structure thoroughly out of touch with its effect on ecosystems (see Bernstein, 1981: 309). Ecological anthropology and associated enterprises therefore are of highly limited explanatory utility in the context of modern industrial societies. Finally, human societies can debate and perhaps even consciously decide aspects of their own structure and behavior; man long ago lost his unselfconscious innocence, and can no longer be numbered just one among the beasts. Unlike the rest of the beasts, man can learn how to live in a variety of niches, and also can create new niches of his own (see Colinvaux, 1978: 212–33).

From here on *I will eschew both the homology of social systems with ecosystems and the treatment of man as just another species within ecosystems.* Human societies possess unique devices for dealing with any disequilibrium in their relationship with biological ecosystems. These devices – such as social choice mechanisms – have no necessary direct counterparts in biological ecosystems. Moreover, their history can be understood in non-ecological terms. However, the fact remains that these devices are society's available means for coping with ecological problems, and so may profitably be analyzed in an ecological light.[2]

The natural trend in the development or succession of ecosystems is in the direction of increased structure and biomass. There is no "natural" trend in the development of human social systems, let alone human social choice mechanisms.[3] The analysis that follows will proceed in terms of "man, the problem-solver" rather than "man, a species like any other." Social choice mechanisms will be treated as man's collective problem-solving devices, which can themselves be chosen and evaluated in terms

[2] The approach I follow has some commonality with what Orlove (1980) refers to as the modern "processual" approach to ecological anthropology. This processual orientation rests on a belief that ecologically functional behavior and structure can be the product of conscious decision-making (Orlove, 1980: 252).

[3] Hegelians, Marxists, and social Darwinists would disagree with me here.

of their capacity to deal with problems. In this chapter I will attempt to specify a set of criteria that any social choice mechanism must meet if it is to cope adequately with ecological problems.

DOES NATURE OR MAN KNOW BEST?

Ecological rationality as a principle and as a form of functional rationality is concerned with low entropy or order in human systems as they combine with natural systems. The human systems of primary interest in this inquiry are social choice mechanisms. Stated in these terms, ecological rationality allows for considerable variation in the relative importance of human systems and ecosystems in producing the good of life support, or low entropy.

One could argue that the entire burden should fall upon human problem-solving mechanisms. Certainly, human beings have the capacity to engage in wholesale ecological engineering. Other species can and do modify their environment to their liking, but man is unique as a species which can create its own niche. It is in this spirit that one speaks of the "management" of natural resources.

At the extreme, a commitment to ecological engineering would yield a totally artificial environment: the replacement of the biosphere by a "noosphere" directed by human minds (see Vernadsky, 1945). A noosphere would be the logical culmination of the idea of human transcendence of nature – the total subordination of the ecosphere to human utilitarian demands through the exercise of instrumental rationality.[4]

However, instrumental rationality in *collective* human decision-making is problematical, as numerous studies of the policy-making process remind us (see, for example, Lindblom, 1965; Pressman and Wildavsky, 1973). Far from pursuing agreed-upon goals in an instrumentally rational and effective manner, modern governments tend either to "muddle through" (Lindblom, 1959), or simply to follow established organizational routines (Allison, 1971): 67–143).

Moreover, present human ecological engineering capacities do not approach those demanded by an effective noosphere. Complexity, uncertainty, and variability in the ecological environment of human choice (see chapter 3) are sufficient to overwhelm any ecological engineering project equivalent to (say) the construction of a building, or an aircraft. Even at the microcosmic level of sustainable, human life-supporting ecosystems of a size that would fit in a spacecraft, technology

---

[4] White (1967) sees transcendence as inherent in the Judeo-Christian religious and ethical tradition.

---

has yet to indicate anything remotely promising (see Spurlock and Modell, 1978). The history of human intervention in ecosystems has its fair share of unanticipated consequences and planning disasters.

Even if wholesale ecological engineering were possible, its desirability would not be thereby guaranteed. For natural systems currently perform productive, protective, and waste-assimilative functions in a spontaneous and unplanned manner; why, then, invest in a costly (and entropic) duplication of those functions (see Hall, 1975; Odum, 1983: 53)? Why build a canal next to a navigable river? One can take advantage of the spontaneous qualities of natural systems, rather than seek to replace them with artificial systems.[5]

At the other extreme from the would-be ecological engineer are those who believe that the spontaneity of ecosystems is best left well alone. So, for example, Barry Commoner (1972: 37) tells us that the third law of ecology is "nature knows best." Acceptance of this "law" implies that we should defer to nature as a superior system designer and maintainer, and dismiss human pretensions to either instrumental or functional ecological rationality as further proof of the "arrogance of humanism" (Ehrenfeld, 1978). Any such argument can provide the empirical backing for an ethical position which Tribe (1974) describes as "immanence" – a total respect for nature and all its works.

There is a sense in which this "nature knows best" contention just has to be correct. Recall that, *in the absence of human interests,* ecological rationality may be recognized in terms of an ecosystem's provision of life support to itself. Left to its own devices, ecological succession tends toward the production of climax ecosystems. A climax ecosystem is one in which there exists a maximum of biomass per unit of available energy flow (Odum, 1983: 445). During succession, negative interactions between species probably tend to give way to positive, symbiotic relationships (Odum, 1983: 369),[6] and resistance stability increases. Succession generally yields increasing homeostasis, and hence increasing ecological rationality. Homeostasis and stability in a mature ecosystem are grounded in functional complexity – that is, in the number of potential feedback loops in the system (see Van Voris et al., 1980) – rather than in structural complexity, or diversity.[7]

[5] One finds this cooperative spirit exemplified in the writings of Aldo Leopold (1933, 1949) on natural resource management.

[6] Some of the most intricate examples of symbiosis can be found in tropical forests and coral reefs.

[7] Tropical forests constitute perhaps the world's most diversified ecosystems, but they are not resilient. The roots of these ecosystems are, quite literally, very shallow. Colinvaux (1978: 199–211) argues that diversity is a consequence of stability, not a cause of it.

There are some minor flaws in this argument on behalf of the spontaneous rationality of natural systems (ask any dinosaur). Most notably, the resilience of natural systems (an ability to recuperate from disturbances) can be expected to decrease with succession (see Odum, 1983: 446). Hence the rationality of an ecosystem may peak at some intermediate stage in succession (as may its degree of functional complexity), reflecting the fact that the maintenance of biomass itself has an energy cost. Optimum carrying capacity – i.e., that sustainable in the face of stress – may therefore be less than maximum carrying capacity (Odum, 1983: 157). However, the capability of natural systems to maintain and develop a functional rationality is clearly well proven.

It is only when one comes to consider human interests in this scheme of things that a "nature knows best" conclusion may be called into question. While the protective and waste-assimilative aspects of anthropocentric ecological rationality may be best served by leaving nature well alone, the productive aspects will not be. Production for human use demands artificial suppression of ecological succession – think, for example, of the artificiality (and extreme instability) of a cornfield.

Ecological rationality in the human interest demands some compromise between productive artificiality and waste-assimilative/ protective "naturalness." Technically, what is required is an "anthropogenic subclimax" (Odum, 1983: 473) – a man-created and man-maintained stable ecological state different from the climax which would obtain in the absence of human intervention.

The possibility of a stable anthropogenic subclimax reflects the fact that low entropy can inhere in the creations of man – in the form of physical capital, human capital (knowledge), social institutions, and culture – and not just in natural systems. Hence one should hesitate before prescribing a minimum of human intervention in ecosystems, and judging as ecologically rational those forms of social choice conducive to such minimization. Indeed, one can imagine combinations of human systems and ecosystems with greater low entropy than ecosystems in isolation. "Nature knows best" entails an excessively dismal outlook on the potential for low entropy in human systems. Human intervention need not be synonymous with ecological destruction.

Ecological rationality suggests, then, that non-intervention in natural systems is untenable. While man can destroy the productive, protective, and waste-assimilative capacities of ecosystems, he is also quite capable of creating and sustaining a stable yet productive anthropogenic subclimax (see Dubos, 1980). Among the best examples here would be the rice paddies of East Asia, often intermingled with patches of forest (see Sears, 1957), or the agro-ecosystems of much of the West European countryside. In the latter case, the landscapes created by Benedictine monasteries are exemplary.

Ecological rationality requires, then, a degree of intervention in natural systems, but falls far short of extreme ecological engineering. Man can make use of rather than seek to supplant the spontaneous self-organizing and self-regulating qualities of natural systems. An ecologically rational man–nature system is one in which human and natural components stand in a symbiotic relationship.[8] Currently, many human actions and human systems are parasitic upon nature (think, for example, of a large city), and, like any unadapted parasite, risk destroying the capacity of their host to support them (Odum, 1983: 401). In the long run, only species that affect their environment positively can survive. Functional rationality in man–nature systems consist of an intelligence that is symbiotic in content. The point is surely that man need not be either nature's master – "transcendence" – or nature's slave – "immanence." Ecological rationality avoids both hubris and sterility.

A SET OF CRITERIA FOR SOCIAL CHOICE

Ecological rationality in social choice may, then, be located in a capacity to produce a symbiotic problem-solving intelligence – low entropy – in conjunction with natural systems. This notion may now be combined with the features of the ecological circumstances of social choice outlined in chapter 3 in the development of a set of criteria to test the claims to ecological rationality of social choice mechanisms, that is, their capacity to resolve ecological problems. The criteria thus developed can then be applied to the evaluation of extant forms of social choice (part II) and the design, should it prove necessary, of novel mechanisms (part III).

Recall that the ecological circumstances of human choice can be captured in terms of complexity, non-reducibility, temporal and spatial variability, uncertainty, collectiveness, and spontaneity. These conditions constitute, in essence, the character of the interactions between human systems and ecosystems as seen from the perspective of human systems. And it is these interactions which need to be governed by symbiotic order if ecological rationality is to be secured.

[8] Technically, that symbiotic relationship need not be an extreme one of "obligate symbiosis" or "mutualism" – a condition in which two communities are totally dependent on one another. Man cannot exist without a supportive nature, but nature can exist without man.

Negative Feedback

That kind of symbiotic order demands, first of all, negative feedback in social choice mechanisms as they interact with ecosystems. Let me stress before elaborating that, in contrast to the use of the term in everyday language, negative feedback here is a *highly desirable* quality! By definition, negative feedback is the presence of deviation-counteracting input within a system. The deviations that one is concerned with minimizing in this context are shortfalls in the life-support capacities of ecosystems in combination with human systems. In principle, stability in a system can be achieved without negative feedback; such, for example, would be a pretension of those who advocate a noosphere, or of the more ambitious central planners. However, in an environment of complexity and uncertainty, one cannot completely understand – let alone plan – that environment, which must (to greater or lesser degree) be treated as a "black box" whose contents are never fully known. As a substitute for perfect understanding of the insides of that "box," any intelligent choice mechanisms will be so structured as to respond to signals emanating from the "box."

In the case of ecological systems, there will be spontaneous homeostatic processes operating within the "box" to which any symbiotically rational social system should be able to react as appropriate. The rationality of the whole – ecosystem plus social system – can be achieved only if negative feedback within that whole system is of high quality. The maintenance of an anthropogenic subclimax requires continual intervention, and therefore continual information about and reaction to the results of intervention. Without negative feedback, an anthropogenic subclimax would lack stability.

Negative feedback can take two major forms. In a teleological system under the control of a central device such as a brain or thermostat, outputs of the system are monitored by that device and deviation-counteracting messages are issued accordingly. In contrast, in non-teleological systems feedback signals do not pass through any central location; control mechanisms themselves are diffused throughout the system (see Patten and Odum, 1981). Ecosystems themselves contain negative feedback only of the diffuse type, but an ecologically rational social choice mechanism could contain either type of feedback device (or both.)[9]

---

[9] Engelberg and Boyarsky (1979) speak of the potential "cybernation" of nature through centralized human controllers. In a similar vein, Deutsch (1977: 26) suggests that collective human decision structures can be the self-conscious part of ecology. These studies both implicitly exclude negative feedback in social choice of the internal and diffuse type.

Negative feedback in a teleological social choice mechanism in combination with an ecosystem is depicted in figure 4.1 The non-teleological
case appears in figure 4.2.[10]

The presence of negative feedback in social choice does not, of itself,
secure ecological rationality across the range of interactions that a
collective choice mechanism will make with natural systems. As long as
feedback instances remain uncoordinated with one another, then
piecemeal social choices, though each apparently rational in isolation,
may fail to produce a rational whole. So, for example, any solution to a
problem in one locality should do more than simply shift the burden to
another area; cleaning up one medium should not cause worse pollution
in another; and a single actor should be able to anticipate that his or her
or its sensitive contribution to the resolution of an ecological problem
will be matched by, and will mesh with, the contributions of the other
actors involved.

Coordination

Thus a second desideratum is required: coordination. Coordination in its
turn is of two kinds. The first is *among* actors but *within* particular
collective actions; the second is *across* different collective actions.

The need for coordination *within* choices stems directly from the
"collective" quality of many ecological problems, described in chapter 3.
Public good underprovision and the tragedy of the commons both
represent a failure of coordination across actors, which can occur even if
negative feedback is adequate. These problems share an underlying cause
which may be represented in the famous "prisoner's dilemma" game (see
Hardin, 1982).[11] Coordination within choices requires some resolution
of the dilemma. The game is diagrammed in figure 4.3 for the simplest
case of two actors who play the game just once.

A and B are the two players of this game. Each has two choices
available – to "cooperate" with the other player, or to be nasty and
"defect" – in a decision both players must make simultaneously. Imagine

---

[10] In passing, it should be noted that not all feedback is negative. Positive feedback, or
deviation amplification, is possible too. Positive feedback exists in natural systems in
processes of growth in organisms and populations, and hence can have value in ecosystem
development. However, that value can be only transitional; if any limiting factors are
approached, positive feedback is a recipe for disaster in the form of "overshoot and
collapse."

[11] The reason for the title of the game is that its usual exposition is in terms of the choices
facing two prisoners who are kept in different cells but are accused of a joint crime: each is
faced with the choice of confessing (and implicating the other) or remaining silent.

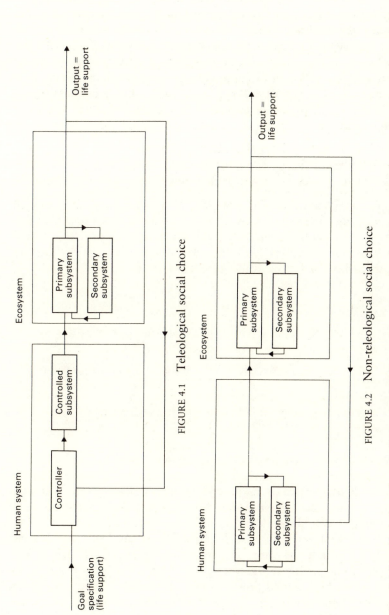

FIGURE 4.1   Teleological social choice

FIGURE 4.2   Non-teleological social choice

Figures 4.1 and 4.2 are based on those in Patten and Odum (1981). In each case, the "secondary subsystem" consists of diffuse, low-energy feedback devices. For the sake of simplicity, only one ecosystem and one human system are depicted. In reality, large numbers of human systems and ecosystems can be connected together.

a situation in which A and B are deciding whether or not to contribute to the provision of clean air in the neighborhood where they both live. Let us assume there exists some common metric for the measurement of both the benefit yielded by clean air and the cost to A and B of making a contribution to its provision. One can then compute a measure of net benefit (or "payoff") to A by subtracting "cost to A" from "benefit to A of cleaner air." This payoff to A appears as the top line in each of the four cells in fig. 4.3. A similar measure can be computed for B in each cell. Assume further that:

The cost of one player's contribution to clean air	= 7 units
The benefit provided by *one* contribution	= 4 units
The benefit provided by *two* contributions	= 10 units

If player A contributes ("cooperates") and player B does not (B "defects"), then the payoff to A is $(4-7) = -3$, while the payoff to B is $(4-0) = 4$. If neither contributes, the payoff is zero to both. If both contribute, the payoff is $(10-7) = 3$ to both. The possibilities are diagrammed in the four cells in figure 4.3.

		Player B	
		Cooperate	Defect
	Cooperate	A = 3 B = 3	A = -3 B = 4
Player A			
	Defect	A = 4 B = -3	A = 0 B = 0

FIGURE 4.3   Prisoner's dilemma game

From A's perspective things look like this. If B defects, then A should defect too, in order to avoid the (-3) loss. On the other hand, if B cooperates, A should *still* defect, in order to achieve a payoff of 4 rather than 3. So, whatever B does, A should defect. From B's perspective the game looks exactly the same. Therefore, both A and B will defect, and they end up in the lower right-hand cell, with a net benefit of zero to each. Clearly, both would prefer to be in the top left-hand cell, where both are 3 units better off. Herein lies the paradox of the prisoner's dilemma: individually rational choices lead to collectively bad outcomes.

This two-person case is clearly an oversimplification of common property and public good collective action problems. A more accurate representation is an "*n*-person" prisoner's dilemma.[12] The *n*-person case differs from the two-person case in that any hurt caused by defection is spread over many players, no one player may know what all the others are doing, and signals to other players are diffuse (see Dawes, 1980). In general, the larger the number of players involved, the greater the temptation to hitch a free ride on the efforts of others, and the less the chance of being caught. Coordination in social choice demands *some* kind of solution to the prisoner's dilemma.

Coordination *across* choices is more readily explained (though no less important) than coordination within choices. The need for coordination across different social choices stems directly from non-reducibility in the ecological circumstances of social choice. The parts of a social choice mechanism must be able to act in concert, such that choices at any point in the mechanism are rationally adapted to choices at other points. So, for example, solution to a pollution problem in one locality should do more than simply shift the burden elsewhere (for example, by building a smokestack tall enough to export pollution hundreds of miles). One actor's solution should constitute more than another actor's problem. Coordination failure of this sort underlies the phenomenon of problem displacement discussed in chapter 2.

Again, as with negative feedback, coordination can be achieved in a centrally controlled (teleological) manner, or through diffuse (non-teleological) processes. Coordination of itself is not, though, a guarantee of ecological rationality in collective choice; the parts of a human system might be coordinated in waging war, or destroying ecosystems. Therefore, negative feedback and coordination must stand together if a social choice mechanism is to be judged ecologically rational.

Robustness or Flexibility

Negative feedback and coordination may be sufficient to cope with uncertainty, complexity, and non-reducibility in the ecological environment of social choice, and to take advantage of spontaneity in that environment. However, these two qualities are insufficient to guarantee ecological rationality in the face of temporal and spatial variability in the ecological circumstances of social choice. That degree of variability demands an additional quality, which may take the form of either robustness or flexibility.

The quality of robustness is the ability of a mechanism to perform well

[12] The *n*-person game can also be depicted in terms of figure 4.3. Instead of standing for one individual, player B represents "all players other than A." The explication changes but slightly (see Hardin, 1982: 22–38).

across a wide variety of conditions. Conversely, adequate performance must not be contingent upon a restrictive set of conditions.[13] Robustness is, in fact, more easily assessed in terms of any such lack of obvious dependence; for one aspect of uncertainty is that one cannot know the full range of conditions across which adequate performance is desired. Pertinent dimensions of conditions might include rate of energy flow through human and natural systems, degree of material prosperity, information availability, knowledgeability and cognitive capacity of the actors in a social choice mechanism, pace of change in the environment, the nature of the forms of social choice with which a given mechanism must coexist, and the content of the problems faced (e.g., productive, protective, and waste-assimilative).

Flexibility is a fairly direct substitute for robustness in social choice. While robustness refers to steadfastness of performance across varying conditions (and varying degrees of stress), flexibility is the capability of a mechanism to adjust its own structural parameters as changing environmental conditions demand. The conditions of interest here are the same as those discussed above under "robustness." A flexible social choice mechanism should, for example, be able to incorporate new kinds of feedback devices as necessary to respond to a novel form of signal, or to establish new forms of coordination to cope with novel emergent problems. Any mechanism with self-perpetuating qualities yielding an automatic resistance to change clearly fails the test here.

Negative feedback, coordination, and robustness *or* flexibility are together sufficient to guarantee the maintenance of the ecological rationality of a social choice mechanism. "Maintenance," though, means just that, and no more. If there exists a state of fundamental disequilibrium in the interactions of human systems with ecosystems, then these four qualities are insufficient to secure the *attainment* of ecological rationality. It is at this juncture that the degree of severity in the ecological problems faced by a social choice mechanism enters the picture. It was pointed out in chapter 2 that severity may be captured in terms of the depletion of low entropy. If human systems are indeed engaged in wholesale depletion of low entropy, then the mere maintenance of a symbiotic equilibrium is not enough.

Resilience

If ecological problems are severe, then by definition stress has succeeded in removing a combination of human and natural systems from its

[13] It should be noted that robustness may mean that a system is suboptimal under any given set of conditions.

Pertinent ecosystem features	The circumstances of ecological problems	Normative judgment		Criteria for social choice structures
Interpenetration	Complexity	Maintenance of the capability of human and natural systems in conjunction to cope with actual or potential shortfall in life support	=	Negative feedback
Emergent properties	Non-reducibility			Coordination: across choices within choices
Self-regulation	Variability: spatial temporal			
Dynamism	Uncertainty			Robustness
	Collectiveness			Flexibility
	Spontaneity			Resilience (contingent upon severe disequilibrium)

FIGURE 4.4 Deriving criteria for social choice

equilibrium range. In the face of severe ecological problems, social choice mechanisms therefore require an ability to steer human and ecological systems back to normal operating range. This quality may be termed *resilience,* after a similar property found in ecosystems.[14] One does not need to be a Malthusian prophet of doom to note some signs of fundamental disequilibrium in the interactions between human and natural systems in the contemporary world (see chapter 2). Hence resilience may be added as the fifth item on a list of desiderata for social choice mechanisms. It should be borne in mind, though, that its status is of a different quality than that of the other four criteria: resilience is a contingent criterion, applicable only in conditions of extreme disequilibrium.

The criteria list for social choice mechanisms now stands complete at five items: negative feedback, coordination, robustness, flexibility, and resilience. Robustness and flexibility are substitutable; resilience is contingent. Hence only negative feedback and coordination are necessary conditions for ecological rationality. These two plus robustness *or* flexibility are sufficient conditions, *unless* one commences from a situation of fundamental disequilibrium, in which case one must add resilience. The derivation of the five criteria is summarized schematically in figure 4.4.

It should be stressed that the five criteria as stated imply no prejudgment of the broad categories of social choice that can meet them. It is reasonable to apply these criteria to mechanisms that are centralized or decentralized, teleological or non-teleological, governed by formal rules or anarchic.

The stage is now set for proceeding with an evaluation of social choice mechanisms in terms of the ecological rationality standard – specifically, in terms of the five criteria established in this section. However, the fact that one *can* apply this standard to the evaluation of social choice mechanisms does not imply that one *should.* In the next chapter I will seek to establish the moral priority of ecological rationality over competing forms of reason. The reader already convinced of that priority may wish to skip directly to part II.

---

[14] However, note that resilience stability in ecosystems refers to the capacity of a system to move *itself* back to the equilibrium range, rather than its ability to so move *other* systems.

5

The Priority of Ecological Rationality

Ecological rationality is something of a novelty. The forms of functional rationality applied and appealed to most often in social systems (and social choice) are economic, social, legal, and political rationality. In this chapter I wish to outline these four competing forms of reason, and to make an argument for the moral priority of ecological rationality over them; for, while all five forms of reason constitute very different and inconsistent guides to practice, they are commensurable, so that their claims can be compared.

ECONOMIC RATIONALITY

Economic rationality is the dominant form of reason in contemporary industrial societies – and, arguably, the defining feature of those societies. Economic rationality is grounded in utility calculations: a system may be judged economically rational if and when net utility is maximized across an aggregation of individuals. Under economic rationality, then, the dominant kind of relationship is based on calculation, and the prime value is economic efficiency.[1]

Most analyses in the social choice idiom proceed in the context of economic rationality: social choice mechanisms are evaluated in terms of rules based on some aggregation of individual preferences. Such rules include axiomatic social welfare functions, real-valued social welfare functions, and more practical principles such as Pareto optimality (see, for example, Mueller, 1979). Indeed, those who undertake social choice analyses often *define* social choice in terms of the aggregation of

[1] Proofs such as those of Bator (1957) attest to the theoretical welfare-maximizing rationality of free markets. Proponents of central planning might dispute such conclusions, but their standard of rationality is identical. Note, for example, the pride taken by Soviet bloc bureaucracies in their production accomplishments.

conflicting individual preferences (a definition inconsistent, of course, with that adopted in this study).

Economic rationality, being ubiquitous, is no stranger to the ecological realm. The interpretation of environmental problems in terms of "externalities" – which have a price, just like any other good, and which need to be corrected if the economic rationality of the system is to be secured – is one example of this mode of thinking (see, for example, Baumol and Oates, 1975). Natural resources, whether renewable or non-renewable, are treated as nothing more or less than inputs to productive processes: factors of production substitutable as efficiency dictates for labor and capital, and fit for the application of a "theory of use" (e.g., Scott, 1973) to determine their optimal rate of depletion. A good social choice mechanism for environmental problems is, in terms of economic rationality, one responsive to the preferences of the individuals directly affected by choices concerning the problem in question – for example, residents of a watershed (see Haefele, 1973).

This form of reason is less than compelling in the ecological sphere, if only because a system may be judged economically rational while simultaneously engaging in the wholesale destruction of nature, or even, ultimately, in the total extinction of the human race. The latter result holds because of the logic of discounting the future (see Page, 1977; Clark, 1974; Dryzek, 1983c: 2). Without discounting, the calculations at the heart of economic rationality become indeterminate. The impetus of economically rational systems is to maximize production; in an ecological context, this goal is in direct conflict with the spontaneous developmental trend in ecosystems, which is protective (see Odum, 1983: 487–8).

SOCIAL RATIONALITY

Social rationality is a form of reason less controversial in the ecological realm than economic rationality, but also less obviously applicable. Under social rationality, the prime value is social harmony and integration, generally achieved through processes of interpersonal adjustment and persuasion which are not mediated by price or numerical calculation. As a standard, social rationality is appealed to by policy-makers, analysts, and advocates far less frequently than economic rationality. No quarrel with social rationality is necessary here, because this form of reason has no obvious application to ecology. A society might be socially rational, yet still be in a state of disequilibrium with its natural environment.

LEGAL RATIONALITY

Under legal rationality, behavior is governed by formal rules. The guiding values are conflict resolution and the construction of a complete system of rights and rules. Again, legal rationality is not, prima facie, undesirable from an ecological standpoint, except inasmuch as it shares the economic sin of treating environmental problems in terms of the conflicting interests of different persons. The rationality or otherwise of legal systems will be discussed at length in chapter 10 below. Suffice it to note for the moment that legal rationality has little to say concerning the substantive content of outcomes of disputes; hence one cannot apply it as a standard sufficient in itself to guide actions in the ecological sphere.

POLITICAL RATIONALITY

The pre-eminent value under political rationality is less easily captured in succinct terms than that under economic, social, and legal rationality. Diesing (1962) describes political rationality in terms of "intelligence," such that a politically rational system is one that can solve the collective problems confronting it. Effectiveness in problem-solving as a value is not easily disputed, but political rationality's claims here rest on Diesing's somewhat idiosyncratic line of argument. In the first instance, political rationality may be recognized by consensus among the relevant – i.e., relatively powerful – political actors. Concomitant with that consensus is a minimization of alienation from (and hence opposition to) the political system in question. Diesing views the consequent high level of support and legitimacy achieved by the system as enhancing its capacity to solve the problems it is likely to face in the future. The difficulty here is that consensus in the context of any one problem may be achieved at the expense of effective problem resolution therein. Summing across all the problems in one system, the result may be considerable support and consensus, but little resolution. Paradoxically, therefore, a lack of effective problem resolution is justified on the grounds of enhancement of problem-solving capabilities. Only if one defines "problem" in terms of the need for mollification of politically powerful interests is this paradox resolved.[2]

[2] Arguments on behalf of political rationality also tend to be indeterminate. The achievement of political rationality in any one instance implies maximal support for the system as it exists; but there is no guarantee that this system is the best of all possible systems (or even a particularly good one), even in terms of a capacity for the promotion of political rationality.

Political rationality is more often advocated as a principle for all the significant decisions made in a political system (e.g., Diesing, 1962; Wildavsky, 1966; Elkin, 1983) than in the context of a specific issue area such as ecological problems. It is clear, though, that political rationality has nothing special to offer in the ecological sphere. If destructive interests have the greatest political clout, then political rationality will reflect those interests. A system can easily be both politically rational and ecologically destructive.

THE PRIMACY OF ECOLOGICAL RATIONALITY

Economic, social, legal, and political rationality all seem less than homologous with effective ecological problem-solving. This recognition adds further backing to the need for ecological rationality as an analytical construct. Ecological rationality is clearly different to these four standard forms; but should it take priority over them?

Diesing (1962: 169–234) and Wildavsky (1966) argue that political rationality should always be the primary concern in any collective decision, for if that decision commands consent then decision structures will gain support and legitimacy. Healthy decision structures are necessary if any problem is to be attacked in the future; political rationality is primary because its achievement is instrumental to the pursuit of any other (economic, social, or legal) goal.

Diesing's argument on behalf of political rationality illustrates the more general principle that a logical antecedent of a value must be accorded at least equal standing to that value. This principle informs arguments such as those of Rawls (1971) on behalf of a number of "primary goods," and Shue (1980) for a set of basic rights. Primary goods and basic rights are of overriding importance because they are the preconditions to a meaningful life.[3]

The preservation and enhancement of the material and ecological basis of society is necessary not only for the functioning of societal forms such as economically, socially, legally, and politically rational structures, but also for action in pursuit of *any* value in the long term. The pursuit of all such values is predicated upon the avoidance of ecological catastrophe. Hence the preservation and promotion of the integrity of the ecological and material underpinning of society – ecological rationality – should

---

[3] For a very different example of the principle of antecedence, see Hart (1955) on the priority of liberty as a natural right. Rawls, Shue, and Hart do not share the fatal flaw in Diesing's argument. That flaw, as noted above, concerns the recognition that the pursuit of political rationality can actually obstruct effective problem-solving.

take priority over competing forms of reason in collective choices with an impact upon that integrity.

A claim of priority for ecological rationality in these terms remains, as yet, a somewhat imprecise guide. Should such priority be absolute, or, at the other extreme, simply a mild presumption in favor of ecological rationality? To what extent can competing forms of reason dilute the claims of ecological rationality?

Any tradeoff between ecological rationality and other forms of reason would involve accepting some risk to life support for a quantity of some other value. In the case of any discrete decision, a case for treating ecological concerns as secondary may often seem reasonable: *my* aerosol can, or *my* country's supersonic transport, will have but an infinitesimal impact on the overall condition of the ozone layer, well worth the benefits to *me*. The point here is that all actors face a like set of incentives: if they all choose to treat ecological concerns as secondary in each discrete decision, then those concerns will be eclipsed. Summing across all decisions, the result will be wholesale ecological irrationality. All actors, quite (instrumentally) rationally, will seek to take a "free ride" upon the efforts of others; that is, they will refuse to contribute to the provision of a public good (environmental quality) from whose benefits they cannot be excluded.

This kind of consideration indicates that the priority of ecological rationality should be *lexical* in character. Lexical priority means that lower values come into play only when designs in pursuit of a higher value are totally complete. A lexical specification may be the only way to prevent actors from taking a "free ride" upon ecological rationality. Without lexical priority, ecological rationality is in danger of annihilation; ecological rationality demands lexical priority if it demands any priority at all.[4]

The merits of lexical priority become somewhat less compelling if one introduces the possibility of uncertainty as to whether an institutional design is making any positive contribution to ecological rationality – or, indeed, uncertainty over whether ecological problems really are very severe (see chapter 2). Considerations of uncertainty do, it seems, dilute

---

[4] The best-known recent appearance of lexical orderings comes in Rawls's (1971) *A Theory of Justice*, where a case is made for the priority of liberty in these terms. Some powerful criticisms of Rawls's treatment have been made (e.g., Barry, 1973; Harsanyi, 1975), mostly on the grounds that it is unlikely that any rational individual would give up an infinitely large amount of wealth for a tiny increase in liberty. The special features of ecological rationality render it immune to these kinds of criticisms, for the priority of liberty does not collapse if it is not lexical, whereas the priority of ecological rationality does collapse in this case.

the claims of ecological rationality to priority, especially lexical priority (cf. Dryzek, 1983c: 14–15). However, lexicality is not crucial to the analysis in the chapters that follow. That analysis stands if ecological rationality is treated as only an overriding principle, rather than a lexically prior one. Indeed, I shall not mention lexical orderings again. The preceding discussion was entered into in order to show just how strong the claims of ecological rationality are. Arguments for the lexical priority of economic, social, legal, or political rationality are barely conceivable, let alone likely to be persuasive.

Part II

Evaluation

6

The World's Social Choice
Mechanisms

A set of criteria for appraising the ecological rationality of social choice mechanisms has been generated in chapter 4. But to what specific entities should we apply these criteria? The world's social choice mechanisms can in fact be categorized in a variety of ways. The most familiar is probably the dichotomy between "government" and "market" choice. This scheme is, to say the least, highly oversimplified. Government itself comes in a variety of stripes, and "market" by no means exhausts the forms of non-governmental collective choice (irrespective of the rhetoric of those who seem to think that shrinking the size of government automatically leads to the market stepping into the vacated areas). Lindblom (1977) classifies systems according to their dominant form of social control. My own classification bears some similarities to that of Lindblom, but I prefer to categorize systems by their predominant form of *coordination*. For what makes a social choice mechanism social (rather than personal) is that the activities of diverse actors are somehow brought into concert. Coordination distinguishes collective choice from individual choice.

The following is a list of methods for coordination in human society. It does not pretend to be an exhaustive list, simply an enumeration of the more prevalent and more striking possibilities:

(a) price signals;
(b) commands promulgated from above;
(c) a set of formal rules to which all actors adhere;
(d) shared norms and behavioral principles;
(e) "partisan mutual adjustment";
(f) formal negotiation over actions to be taken by each party;
(g) force;
(h) conditional cooperation;
(i) discussion about appropriate values, principles for behavior, and actions.

Some of these methods need no explanation; others may be a little less familiar. Coordination through price signals represents Adam Smith's "invisible hand." Item (b) requires, and item (d) can feature, coordination by a leader or leadership group, though the type of leadership exercised differs substantially between the two. Item (c) achieves coordination through adherence to a fixed set of rules, rather than variable commands. Item (e) requires a little more explanation. The term is Lindblom's (1965), referring to the process whereby A acts in response to the context for decision provided by the actions taken (or indicated) by B. Nobody is commanding A to make this adjustment; no formal rules say he must do so, no money changes hands, and A and B can adhere to different values. All that occurs is a calculation by A as to what behavior may best promote A's interests in the context now changed by B's actions. This process is nothing more than the give and take of political interaction.

Item (f) – formal negotiation – requires that the parties to a collective choice attempt to resolve their differences through the making of offers and counter-offers. Item (g) is simply the prevalence of the will of the strongest actor, where strength is measured in terms of the capacity to effect physical harm. Coordination is achieved in a trial of strength. Item (h) occurs when self-interested actors make their commitment to mutually beneficial actions contingent upon a like commitment by the other actors in the system (in some senses, it is a sub-category of item (e)). The term "conditional cooperation" has been developed in the context of game theory. Item (i) involves attempts to reason through (rather than negotiate) a consensus on which collective actions may be based.

Each of the nine modes of coordination defines a social choice mechanism, as shown in table 6.1. Some of the forms of social choice listed in the table are familiar, others less so. Rather than go into detail

TABLE 6.1 Defining social choice mechanisms

Coordination mode	Social choice mechanism
1 Price signals	Market
2 Command	Administered system
3 Formal rules	Law
4 Promulgated values	Moral persuasion
5 Partisan mutual adjustment	Polyarchy
6 Formal negotiation	Bargaining
7 Force	Armed conflict
8 Conditional cooperation	"Radical decentralization"
9 Discussion	"Practical reason"

here, let me refer readers to the relevant chapter in parts II and III of this study for a discussion of the finer points of each kind of social choice. Items 1–5 each have a chapter named after them in part II. Items 6 and 7 both appear in the chapter on the international system. Items 8 and 9 are probably the least familiar to a majority of readers; they constitute my own agenda for reform, articulated in part III of this study.[1]

The taxonomy of social choice mechanisms developed here does not pretend to be collectively exhaustive and mutually exclusive. I would defend it on the grounds that (a) it is fairly parsimonious, (b) most real-world cases of social choice can be located within it,- and (c) intelligent analysts such as Dahl and Lindblom (1953) and Lindblom (1977) use similar (if less extensive) conceptual schemes. Ultimately, the true test of the validity of this taxonomy lies in the power and plausibility of the analyses based upon it – in other words, in the entirety of this book.

The social choice mechanisms I have outlined are not equally important. As far as *what actually exists* is concerned, markets and administered systems probably have the strongest presence in today's world. I say "probably" because there is no good empirical measure of the relative frequency of modes of social choice. (Ideally, I suppose one would like an indicator of the sort "proportion of important collective problems approached through mechanism X.") Locally, other forms may be prominent; for example, in Western industrial societies social choice through polyarchy may be as frequent as choice through administration.

One finds a somewhat different mix when it comes to what people think *should* exist. In the USA and the UK, contemporary discourse over ecological problems – and, for that matter, policy problems more generally – has its tone set by proponents of market or quasi-market mechanisms. This situation is a far cry from that obtaining a mere ten or fifteen years ago, when purposive government action was widely regarded as the proper antidote to the ecological abuses attendant upon the unregulated private enterprise system. However, communities of advocates can still be located for centrally planned administered systems, polyarchy, law, and moral persuasion as the appropriate means for dealing with ecological problems, *especially* among those who write, advise, or agitate about these problems. This study will attempt to recognize and respond to the arguments of these various intellectual

---

[1] In basing this taxonomy on modes of coordination, I have used coordination as an *empirical* standard for the *classification* of social choice mechanisms. In chapter 4, the notion of coordination was developed as a *normative* standard for the *evaluation* of mechanisms.

communities, while bearing in mind that it is a legitimate ambition for policy-oriented study to change the parameters of discourse, rather than simply staying within the existing agenda. I believe – and will seek to show – that all the communities I have outlined above are in substantial error.

The remainder of part II will assess in an ecological light the capabilities of these major existing mechanisms. I shall begin with the market.

7

Markets

MARKET RHETORIC

One of the more striking shifts in the tone of political discourse in recent years is the rediscovery of the free market. While celebrations of the virtues of the market go back to the days of Adam Smith, the near-fatal dent in the market's reputation left by the great depression forced many market proponents to go underground, languish on the fringe of political discourse, or move to the University of Chicago.[1] The contemporary resurgence of market thinking has its world headquarters in the USA, where conservative "think-tanks" beyond count churn out their analyses. Popularizers of market ideas peddle their wares in magazines and newspapers. The movement has a thriving outpost in the UK, where "Victorian values" are the order of the day in some circles. However, it is in the USA that market advocates have most firmly seized the intellectual high ground and set the terms of political discourse. Market strategies are the new public policy orthodoxy.

But what does this intellectual ferment have to do with ecology? If we are to have more of the market around, then presumably its ecological effects are worthy of scrutiny. Such investigation has rarely been undertaken by contemporary marketeers, who generally believe that the economy is far more important than ecology. In the terms outlined in chapter 5, they stress economic rationality to the exclusion of ecological rationality; and they see the market as the best means to economic rationality. In the obscure language of the 1980 national Republican Party platform, "the economic prosperity of our people is a fundamental part of our environment."

Many of the marketeers regarded environmental regulation as an

---

[1] A "Chicago School" of economics kept the market flame alive for several decades following the 1930s depression.

especially nefarious impediment to industrial efficiency and growth. Thus Weidenbaum (1983) suggested that the goal of environmental policy in the 1980s should be to "free the Fortune 500" from the burden of excessive regulation. An enthusiastic administrator of the US Environmental Protection Agency – Ann Gorsuch, installed in 1981 – concurred. Gorsuch saw her mission in terms of facilitating market-based economic recovery, not environmental protection.

Leadership of like motivation took over the US Department of the Interior at the same time. Secretary James Watt saw his prime responsibility as expediting the transfer of public land into the private sector, in order to promote the productive use of energy, mineral, timber, and grazing resources in the market economy. Watt and his president (Reagan) were both self-confessed "sagebrush rebels," identifying themselves with a movement to "open up" public lands for commercial exploitation.

A third target of the marketeers in the Reagan administration (ultimately frustrated by Congress) was abolition of the US Department of Energy. Henceforth, energy would be produced and distributed by an unfettered market (with the exception of nuclear energy, for which public subsidy would continue).

Fiasco and scandal in the Watt–Gorsuch years in Washington (see Vig and Kraft, 1984, for a chronicle) tended to obscure the merits of the market agenda as applied to questions ecological. As noted above, this agenda tends to regard ecological issues as secondary to economic concerns. However, inasmuch as the market agenda deals directly with ecological questions, its main planks are as follows.

Transfer ownership and control of public natural resources into private hands. Extend the scope of private property rights to resources traditionally held in common, including air and water.

Remove all regulations on the prices of natural resources.

Eliminate government subsidy for resource use (for example, public projects to enhance the development potential of barrier islands and flood plains).

Rely as much as possible on private sector "voluntarism" to control pollution, with no government sanction or coercion.

If voluntarism fails, devise public policies which mimic market incentives to the greatest extent possible (for example, use a system of standards and charges rather than regulation for pollution control).

Some elements of this agenda did indeed receive attention from the US federal government in the early 1980s. Energy price deregulation proceeded apace. The USA withdrew its assent to the United Nations

Conference on the Law of the Sea accords on the grounds that they provided for public rather than private exploitation of ocean-floor mineral resources. The hazardous waste programs of the Environmental Protection Agency (EPA) would, according to its assistant administrator in charge of the Office of Waste Programs Enforcement (Rita Lavelle), rely on the producers of toxic wastes "assuming their responsibilities and bringing about timely resolution of the problems" (quoted in Cohen, 1984: 285). Thus the EPA ceased to take action against violators of hazardous waste regulations.

However, for the most part market rhetoric was not translated into policy practice. Moreover, many of the pro-market moves that *were* taken in the early 1980s met sharp reversal after the departure from office of their more zealous advocates. But this failure tells us more about the way American government works than it does about the merits or shortcomings of the market in solving ecological problems. The best evidence about the ecological strengths and weaknesses of the market comes not from a few abortive American public policies, but from the actual workings of the market over recent decades and centuries. So just how good is the market?

THE CAPABILITIES OF MARKET SYSTEMS

A market is a means of social organization defined by free and open material exchange among its participants. Markets of varying degrees of freedom and openness are, of course, ubiquitous in today's world. One can debate the historical reasons for this prevalence, but clearly markets have the ability to produce collective outcomes at a "transactions cost" far lower than that of any other social choice mechanism. Moreover, the outcomes produced by markets have several very attractive properties. But the real beauty of markets is that these properties may be realized without any steering of the system toward them. The outcomes are, instead, a byproduct of base calculations of individual self-interest, guided only by Adam Smith's "invisible hand." Market systems, like ecosystems, are non-teleological and self-correcting.

If a market system is working tolerably well, its desirable products will include allocative efficiency,[2] coordination of the activities of vast numbers of producers to common purposes, continuous incentives to

---

[2] Efficiency constitutes an equilibrium at which the net preference satisfaction of an aggregate of consumers is maximized, subject to the constraint of available resources. See Bator (1957) for the classic proof of the efficiency of free competitive markets.

innovate, and (arguably) liberty.[3] What, though, of the ecological rationality or otherwise of market systems? A case can be made that markets do indeed possess the qualities demanded of ecologically rational social choice mechanisms.

Consider first of all the negative feedback requirement. Feedback in market systems exists in the diffuse and internal controls of price signals. The price system constitutes, in effect, the "secondary subsystem" in which the negative feedback devices of a non-teleological, self-regulating system inhere. If a good becomes more scarce, then the normal workings of the forces of supply and demand will cause its price to rise. This price increase will in turn elicit four further reactions: a search for new sources of the good in question, the exploitation of previously sub-marginal supplies, attempts to develop substitutes for the good, and more efficient use of the good. This logic can be applied directly to the goods called "natural resources" (Rosenberg, 1973). So, for example, a scarcity of wood, the key fuel in sixteenth- and seventeenth-century Europe, led directly to the development of coal as a fuel.[4] More recently, increasing energy prices stimulated a far-flung search for new deposits of familiar fuels such as oil and gas; research and development into new technologies such as fusion, solar energy, and synthetic fuels; and conservation.

Thus, it is hard to point to any resource which has ever "run out" in any meaningful sense in a contemporary market system; substitutes have always been developed, and human life support – the "output" in which one is interested here – has been maintained. Indeed, the long-term trend in modern market systems has been one of declining real prices of natural resources (interrupted by an upward turn of energy prices in the 1970s) (see Barnett and Morse, 1963). Julian Simon (1981: 15–27) interprets this evidence to mean that natural resources are becoming *less* scarce with time.

The negative feedback capabilities of markets are additionally attractive inasmuch as markets offer full rein to the forces of innovation and ingenuity – which can constitute part of the "response" side of the feedback cycle. Markets can both promote feelings of personal autonomy

[3] There are those who argue that the freedom of material exchange demanded by markets is a necessary condition for individual liberty in the political realm (Hayek, 1944; Friedman and Friedman, 1962). On the other hand, it may be that market freedoms and political freedoms are both products of one set of historical circumstances, which Lindblom (1977: 162) calls "constitutional liberalism." For a discussion of the implications of "liberty" in social choice from an ecological perspective, see chapter 9 below.

[4] Arguably, this development in turn made possible the innovations at the heart of the industrial revolution (see Nef, 1977).

and effectiveness, and enhance the information-processing capacities of individuals (see Lane, 1978). Market man is an efficacious problem-solver. Ecological problems may be just another category ripe for the "technological fixes" which market systems are so good at providing in response to feedback through the price system.

Negative feedback through the price system can, arguably, also cope with problems of environmental pollution, provided only that there exists a well-defined set of private property rights. If polluter A is harming individual B, then B, even in the absence of any liability rule, can offer to pay A to cut back on his pollution. That offer constitutes, in effect, a feedback signal; for the more harm B is experiencing, the more she will be willing to pay A to cut back; and the more A is paid, the more he will be willing to cut back (Coase, 1960). Therefore demand and supply schedules exist for pollution abatement, just as they do for any good; and a free market will ensure that the economically optimal amount of abatement is produced.[5]

In addition to its negative feedback quality, the price system acts as a coordinating device. Complex problems of allocation, distribution, and patterns of resource use are "solved" by the "invisible hand." Coordination through markets can be achieved at any level of social choice, from local to global.

In addition to its more traditional set of virtues, then, the unfettered market possesses, in negative feedback and coordination, some claims to ecological rationality. Moreover, markets are robust; they can and do function across a wide variety of circumstances. Flexibility and resilience are less readily located in markets; however, given robustness, flexibility may be forgone with little loss, and resilience is but a contingent criterion (see chapter 4). So markets have their attractions in ecological terms.

MARKET FAILURES

Market failings in the ecological realm are often discussed in terms of what economists call "externalities" – unintended consequences of actions which affect, for better or for worse, the production functions of

---

[5] This "Coasian" formulation is easily criticized, and seems totally inapplicable to any conceivable real-world situation (Randall, 1974). Nevertheless, given some *very* strong assumptions, it remains compelling as a theory, and has a number of adherents among free market economists. It should be remembered too, though, that "optimality" is defined here, as always in economic analysis, in terms of the utility and production functions of the parties involved; in other words, the optimum is economically rather than ecologically rational.

other enterprises or the consumption functions of other individuals.[6] The classic example of an externality is environmental pollution: the harm done by local residents' inhalation of noxious emissions from a factory is not reflected in the factory's balance sheet. The concept of "externality" is, however, rooted in an economic conception of the world, not an ecological one. So it will not be emphasized here. The shortcomings of markets will instead be examined in terms of the integrity of the natural systems upon which human existence depends.

Positive Feedback: The Logic of Economic Growth

While markets, as pointed out in the previous section, possess a number of negative feedback devices, they also incorporate a significant amount of *positive* feedback – that is, deviation-enhancing input. Let me stress here that positive feedback is an *undesirable* quality, contrary to the ordinary-language use of the term. Positive feedback in the market is internal to the social choice mechanism itself; feedback signals do not cross the boundary between market and ecosystems.

To put it bluntly, markets are addicted to economic growth. While it is true that economic growth can occur at equally rapid rates in other types of economic system, such as centrally planned economies, market systems are unique in possessing an acute need for continuous growth. Non-existent or low growth in a market system means low investment rates, which in turn lead to high unemployment. Moreover, unemployment will continue to rise insofar as productivity gains are still realizable. Markets need growth if this dismal prospect is to be avoided.

An absence of growth threatens the foundations of market systems for another reason. Inequality in the distribution of income is central to the operation of market systems, as it provides the appropriate structure of material incentives. Under static conditions, inequality is a recipe for discontent. A growing economy provides regular increments to national income, part of which can be directed to the poorer members of society. Thus the absolute income of the poor increases, though inequality may remain unchanged. Economic growth is therefore a device which enables society to avoid confronting tough questions of distribution. Conflict over the distribution of the economic pie is a far more pressing issue in a low-growth, "zero-sum society" (Thurow, 1980).

Modern governments in market systems therefore quite reasonably

[6] It has long been noted that there exists a class of "social costs" which do not enter into the calculus of actors with an eye on monetary return (see, for example, Kapp, 1971). The only costs that enter into the calculations of actors within market systems are private costs – that is, costs whose incidence is felt solely by the actor making the decision.

fear any cessation of growth beyond short-term recession. The members of such governments try desperately – and, indeed, usually see it as their primary task – to ensure the continuation (or resumption) of economic growth.

The profit motive, central to the operation of market systems, acts as an engine of economic growth. But that motive also promotes another positive feedback loop. In order to prosper, producers in market systems must persuade consumers to buy products in ever-increasing quantities. To this end, market systems foster all the seven deadly sins, with the exception of sloth, for "foul is useful and fair is not" (John Maynard Keynes, quoted in Schumacher, 1973: 24). Hence markets are not mere neutral devices for preference aggregation; any predominance of market social choice in a society is likely to have a significant effect on the content of those preferences. Hedonism is fueled in the more well-developed cases by a barrage of advertising.[7] The ensuing culture embodies material acquisition as the highest purpose in life. If people ever did come to believe they had enough material goods, then the market would crumble.

Indefinite exponential growth in a finite world is a logical impossibility. It is on this very simple Malthusian recognition of a finite resource base that systems models of doom (for example, Meadows et al., 1972) are based. One does not need sophisticated computer models to predict the results.[8]

Friends of the market argue that no such finite ceiling exists; the price system ensures that as resources become more scarce then substitutes will be developed (see above). Hence negative feedback will be sufficient to overcome any problems caused by the positive feedback of growth. According to this view, the fact that specific resources are finite presents no insurmountable problem.

However, as argued in chapter 2, there does exist one real finite limiting factor in this world: low entropy, the resource for which there is no substitute. In the presence of this absolute thermodynamic scarcity, Kenneth Boulding notes that "anyone who believes that exponential growth can go on forever is either a madman or an economist." At some point, human beings will need to start thinking about ways to live

[7] Bell (1976) explores the hedonistic requirements of capitalist market systems, and notes that hedonism is inconsistent with capitalism's other cultural requirement, a work ethic. Bell believes that this cultural contradiction may eventually lead to capitalism's undoing if it is not rectified.

[8] Some observers see social and political limits to growth being hit long before any physical or ecological ceiling; see, for example, Miles (1976).

without exponential material growth, and, perhaps, without the market which demands it.

Negative Feedback Inhibition: Myopia and Market Morality

The power of positive feedback in market social choice imposes an extra burden on the negative feedback devices discussed in the previous section. Unfortunately, negative feedback in the free market is itself weakened by two factors: myopia, and market morality.

Myopia here is an inability to respond to signals concerning potential shortfalls in the life-support capacity of ecosystems beyond the immediate future. It is not true that markets *neglect* the future: if an actor within a market system believes that a resource will become more scarce in the future, then it is economically rational for that actor to invest in ownership of the resource in question and to withhold it from the market in the expectation of a future increase in price. "Futures" markets are well-developed, and prices in those markets might be expected to constitute vital negative feedback (or "feedforward") signals.

The flaw here is that any investment in the future has an opportunity cost: market rates of interest. Thus, economically rational actors must discount expected future benefits (and costs) by the rate of interest expected to prevail. Any positive rate of interest therefore shortens the time horizon of actors within the marketplace, and of the market as a whole. The higher the rate of interest, the more myopic the system becomes. At a rate of 10 percent, a dollar's worth of benefit or cost expected to be felt in 20 years' time is valued at 15 cents today. The major consequence is that any effects of present choices which will be felt in the moderate to distant future are accorded little weight in market social choices. The market facilitates the kind of temporal problem displacement outlined in chapter 2. This generalization holds for potential catastrophes, such as total depletion of a "limiting" resource (e.g., fossil fuels), or even wholesale ecological collapse.

If ecological resources were indeed infinitely substitutable for one another, then discounting the future at market rates of interest would produce no ill-effects. Indeed, it can be demonstrated that an optimal time-profile of resource use would result (Scott, 1973). But if there are limits – such as the thermodynamic limit of low entropy – markets will yield no anticipation.

A market system without interest rates and discounting – and therefore without negative feedback inhibition – is not easily envisaged. Individual actors would have to consider infinite streams of costs and benefits in their buying, selling, and investment decisions – an impossible burden of calculation. Further, interest rates represent the opportunity cost of capital; in their absence, capital would be essentially free. Borrowing money would be costless. Interest rates occupy a central place in market systems; it is hard to imagine a market functioning without them; and as long as they remain, myopia will be a significant factor inhibiting negative feedback.

Negative feedback signals from natural systems are further inhibited by the miserable morality of the marketplace. The culture of material acquisition (noted above) equates happiness with income. Areas of life involving values other than material income and consumption are therefore downgraded (Lane, 1978). At best, those values will be converted into monetary terms. Price is the dominant measure of value in the market; environmental resources, like any other commodity, have a price, and so can be traded off against other commodities. "The environment" is treated in terms of raw materials in the production functions of enterprises, or as amenity in the utility functions of individuals. Economic rationality reigns supreme. The discipline of economics can assist here in the computation of "shadow prices" for environmental values (McKean, 1968). Such calculations will not, in general, interest market actors, who care only about exchange values. Only non-market actors – especially governmental ones – will pay attention to shadow prices. But irrespective of any audience, calculations by producers, consumers, or economists in market systems downweight the less tangible protective and waste-assimilative aspects of ecosystems in comparison with productive values more readily expressed in terms of price.

Coordination Problems: Common Property and Public Goods

The price system of market mechanisms can be an effective decentralized coordinating device, such that economically – and perhaps even ecologically – rational outcomes can be attained at the collective level. However, market coordination breaks down in the cases of common property and public goods (which were described in chapter 3).

The "tragedy of the commons" has been extensively documented for a number of resources. For example, the history of ocean fisheries under systems of free access has been one of overfishing, overcapitalization, depletion of fish stocks, and declining incomes (Gordon, 1954).[9] On a larger scale, the increased rate of growth in ecological problems in recent

[9] More recently, systems of free access in ocean fisheries have in some cases been replaced by government-supervised regulatory systems, which have had some successes in ameliorating the tragedy of the commons. Such solutions, however, involve stepping outside market social choice.

Evaluation

decades can be interpreted in terms of the interaction of market incentives and common property. According to Commoner (1972), profit-motivated technological change in the industrial world since 1945 has generally accomplished a shift of production costs away from those borne privately by productive enterprises and toward those imposed upon the "commons" of the environment.

Common property problems are not unique to market systems; they can arise in any decentralized decision system populated by economically rational actors in which rights are not matched by equivalent responsibilities. The market is only one example of such a system. However, the free exchange at the heart of the market all too easily translates into free access to commons resources.

The lesson of the tragedy of the commons is that "freedom in a commons brings ruin to all" (Hardin, 1968) – or, more specifically, that freedom in a commons *under a system governed by unrestrained private pursuit of self-interest* brings ruin to all. This lesson is far from universal acceptance. So, for example, the stance of the US government toward the United Nations Conference on the Law of the Sea is that an international commons should be *created* for the mineral resources of the ocean floor. This attitude stems, arguably, from traditional American beliefs about the virtues of unfettered free enterprise. The so-called "sagebrush rebellion" in the American West in the early 1980s was in large measure an attempt by cattle ranchers and others to accentuate the "commons" aspects of public lands.

The solution to commons problems advocated by most friends of the market is the introduction of a structure of private property rights (Pejovich, 1972; Stroup and Baden, 1983). To extend the commons metaphor, the answer to overgrazing on common land is to divide that land into enclosed plots and grant exclusive title to each of them. The private owners of these plots then have every incentive to careful husbandry. A modified version of this approach can even be applied to the waste-assimilative aspects of natural systems; some public authority can establish the assimilative capacity of (say) a lake, and proceed to auction off rights to discharge specified quantities of pollutants into the lake (Dales, 1968). In schemes of this sort, the coordination problem disappears along with the commons itself. However, the introduction of any new system of property rights requires an authority capable of changing the structure of rights and enforcing the new structure.[10] The

---

[10] It is true that any market presupposes the existence of an authority to guarantee private property rights. It is when one actively seeks to change the structure of rights that the demands upon the authority multiply.

presence of such an authority removes us from the realm of relatively pure markets into mixed systems, which will be discussed in more detail below.[11]

Theories of coordination in the free market apply only insofar as the production of *private* goods – those that can be bought and sold, used and enjoyed in discrete packages – are concerned. Any claim to coordination is inapplicable in the case of *public* goods. In any system of free exchange, public goods either will not be supplied or, at best, will be under-supplied. The reason for this is that there is no incentive for any actor with an eye on financial return to contribute to the provision of such goods.

In practice, *pure* public goods are rare; a class of quasi-public goods with elements of privateness is larger.[12] However, it is clear that several of the environmental public goods listed in chapter 3 (for example, biotic diversity and low entropy) are relatively pure cases. It is hard to imagine a market in them. Market coordination breaks down in these cases. This recognition of the public good nature of ecological integrity means that the Coasian analysis of environmental problems referred to above cannot work. For, typically, in any case of environmental pollution, the number of victims is very large; in attempting to induce polluters to abate their emissions, these individuals would be trying to supply themselves with a public good. Negative feedback may be highly pronounced inasmuch as these individuals are choking on pollution and willing to pay for its abatement, but their need to supply themselves with a public good will inhibit the coordination which ecological rationality demands.

It was noted in chapter 4 that overcoming the coordination problems arising in the context of common property and public goods requires some solution to the "prisoner's dilemma." Markets inhibit any such solution because they promote the materialistic, avaricious, and competitive side of human nature, and hence the likelihood that individuals will choose to "defect" rather than "cooperate." This tendency will be

[11] An exception to this requirement may occur if there is a spontaneous process, analogous to an "invisible hand," that acts to define private property rights. Demsetz (1967) and Anderson and Hill (1975) see such a process occurring as a common property resource becomes more heavily exploited; their examples, though, refer solely to resources such as land, flowing water, cattle, and fur-bearing mammals which may readily be designated in terms of private property rights.

[12] So, for example, the good of environmental quality can, to a degree, be bought and sold in markets; individuals can purchase a cleaner environment by moving from the inner city to a more expensive house in the suburbs. Property values in a market system do increase with air quality (see Ridker and Henning, 1971). "Solution" to the coordination problem through such purchases is analogous to solving commons problems through privatization.

especially acute if the individual recognizes the market-determined character of the other player(s) in the game.

However, a more important inhibition to cooperative resolution of the prisoner's dilemma arises inasmuch as the market places individuals in situations in which they interact infrequently with any other given individual. Axelrod (1984: 129) notes that cooperative behavior in the prisoner's dilemma is a positive function of the frequency with which players encounter one another (because conditionally cooperative strategies can then be tested and chosen). Market interaction is sometimes recurrent, between particular buyers and sellers, or workers and their employers. Thus market actors in Japan especially seek bonding and frequent contact with one another. However, the purer the market, the more infrequent and anonymous is social interaction. Any one individual's transactions are spread across large numbers of other people. The more competitive the market, the greater the number of buyers and sellers, and the less the likelihood that any two of them will repeat an interaction. One goes where the price is best, rather than where one went last time.

Robustness, Flexibility, Resilience

Overall, it is clear that competitive, private enterprise market systems fall short on the two necessary conditions for ecological rationality: negative feedback and coordination. Negative feedback devices do exist, but they are weakened by myopia and market morality; moreover, those devices can expect competition from the positive feedback central to processes of economic growth. "Invisible hand" coordination breaks down under circumstances of common property and public goods.

These failures on the grounds of negative feedback and coordination make questions of robustness and flexibility somewhat moot. For what it is worth, though, markets do possess a high degree of robustness (see above). This robustness cannot, however, refer to performance: given their above-noted failings, markets can be expected to perform poorly across a wide range of contexts.

Markets lack any clear flexible qualities. While at first sight it might seem that they can adapt quickly to new circumstances through changes in the mix of goods and services they produce and the concomitant patterns of distribution and resource use, the central elements of market structure are highly resistant to change. Those elements include the nature of exchange relationships, private property, interest rates, and the positive feedback of economic growth. There are no internal control mechanisms which could change those features; any such alteration would therefore have to come about through the application of a very visible external hand. Historically, structural changes have been a product of the hand of government; indeed, the very dominance of market social choice in Western countries was itself achieved through active government (Polanyi, 1944).

The paucity of negative feedback and coordination in market systems might perhaps be forgiven if markets constituted "pioneer" forms of social choice, with an internal logic that paved the way for more "mature" types capable of sustaining a symbiotic relationship with the natural world. There is a sense in which markets – at least under capitalism – might perform this function, if one accepts Marxist historicism. However, if markets were resilient in this manner it would only be through their own collapse. One should hesitate before ascribing ecological rationality to markets on the slender hope that their collapse might somehow herald the inception of a novel, more benign, kind of collective choice.

CAN MARKETS BE MODIFIED?

The analysis of this chapter to date has been directed at markets characterized by consumer sovereignty, self-interested motivation, private enterprise, and a high degree of competitiveness among buyers and sellers. Markets can and do exist, though, which lack one or more of these features; indeed, the number of possible permutations is considerable.[13] Rather than attempt an exhaustive search of every conceivable variety of market system (an impossible task), in this section I will try to identify the aspect(s) of market-type systems responsible for the failings in negative feedback and coordination discussed above, and to investigate the prospects for purging the guilty aspect.

Positive Feedback: Growing Pains

The necessity for growth in market systems arises from a combination of the profit motive and private enterprise. Under these circumstances, market systems have inbuilt positive feedback that produces economic growth. While it is true that market systems can exist (and have) in the absence of growth, such a state of affairs persists at the peril of ever-increasing unemployment. One partial "cure" for growth might be to increase the degree of monopoly and oligopoly in a market system, for

---

[13] See Lindblom (1977: 93–106) for an outline of a number of variations.

production is (according to economic theory) lower under such circumstances than it is under competitive conditions. Alternatively, the prospects for public enterprise as the foundation for a market system might be explored. Public enterprise will be discussed in detail below.

Incurable Myopia

The myopia of market systems which inhibits negative feedback has no cure, for in the absence of interest rates a number of goods would be effectively free. Markets in capital simply cannot exist without interest rates.[14]

Modifying Market Morality

The kinds of negative feedback inhibiting values associated with market systems result from private enterprise in combination with the profit motive, to which the comments of the above paragraph on positive feedback are applicable. Aside from public enterprise, the only way to adjust market morality would be to graft some kind of value-adjusting device onto the market. Any such device would require moral persuasion through education, discussion, criticism, or indoctrination. Moral persuasion as a form of social choice will be treated at length in chapter 11 below. Persuasion of this sort would run afoul of the constant efforts of producers to persuade consumers to purchase products. Hence effective moral persuasion would also seem to demand the elimination – or at least a radical reduction – of the role of *private* enterprise.

Common Property Resources Revisited

The coordination problem of the tragedy of the commons results from a large number of self-interested actors pursuing their advantage under a particular structure of property rights. Candidate solutions to the tragedy are therefore (a) change the motivation of actors; (b) reduce the number of actors having or controlling access to any particular resource; and (c) change the structure of property rights – that is, institute a system of private, public, or communal property.[15]

---

[14] Iran's Islamic Republic at one point had leanings in the direction of abolishing interest rates on the grounds of Islamic prohibitions against usury. The most likely result of such a move would be that interest rates would get a different name.

[15] Public property vests ownership in the state, which then exercises monopoly control over the resource. So, for example, large tracts of forest in the USA and UK are publicly owned. Under a system of communal property, ownership is vested in the community of resource users, who henceforth make decisions about access to the resource on a collective basis.

Option (a) would require effective moral persuasion, such that enlightened and public-spirited individuals would spontaneously avoid over-exploitation of common property resources. This consideration brings to mind Edmund Burke's dictum that "Men are qualified for civil liberty in exact proportion to their disposition to put moral chains on their own appetites," such that "men of intemperate minds cannot be free."[16]

Option (b) – a reduction in the number of actors having access to a resource – can be accomplished through control on the part of some central authority: a government. Governments can choose to regulate the nature of access to a resource; to actually reduce the number of people entitled to access; or to act as monopoly public authorities themselves.

Option (c) – changing the structure of property rights – again requires governmental action if the shift to a new set of rights is to be effected and enforced. In the case of global or international common property resources, this governing authority must be of corresponding scope. Any system of public property requires administrative social choice, and communal property is one defining feature of anarchism. Administered and anarchistic social choice will be dealt with, respectively, in chapters 8 and 16 below.

The Provision of Public Goods

Public goods underprovision is a consequence of much the same set of factors as commons problems, though the structure of property rights is less important. The above comments about the potential for change in the content of individual motivations therefore apply to public goods too. A reduction in the number of actors involved in a system is also a conceivable solution. Any such reduction would increase the frequency of interactions among particular individuals, thereby promoting the likelihood of cooperative outcomes to prisoner's dilemma problems. Reducing the size of market systems such that they would operate at only a very local level is one possibility here, if a fairly drastic one. Another very direct way of reducing the number of actors in a system is to establish a public authority to provide public goods. Aside from the obvious point that such an authority is itself a public good, though, governments themselves are not immune to free-rider "defections" from cooperative solutions.

---

[16] Ophuls (1977) considers this statement so important to the spirit of his Malthusian opus that it appears as the frontispiece to the book.

MODIFIED MARKETS

The modifications to market systems suggested in the previous section fall into three main categories: a change in the structure of markets, the introduction of some kind of moral persuasion, and an increase in governmental control. These three possibilities will be discussed in detail in this section.

Structural Change

The structural changes mentioned in the last section are of two very different types. First of all, a shift from competition toward monopoly – be it public or private – might both alleviate pressure on common property resources and curb the positive feedback of economic growth. Such a shift would constitute, in essence, a movement from diffuse to centralized coordination. Monopoly necessitates, though, administered rather than market social choice. Administered systems will receive a full treatment in chapter 8 below.

A very different kind of structural change would involve attacking commons and public goods coordination problems through a reduction in the number of individuals (or other actors) in a market system. Any such reduction would necessitate a constriction of the geographical scope of market exchanges. The ensuing constricted markets would operate among a relatively small number of people in a well-defined area. Incentives to "defect" by exploiting common property or providing public goods would thereby fall. A considerable degree of autarky would be necessary to prevent this kind of system from slipping back toward large-scale markets. One might even explore the possibility of defining the scope of each economic unit by ecological boundaries. This kind of localized autarky in which the market prevails appears close to the kind of world which Lovins (1977) envisions should society have the good sense to choose his soft energy path. However, restricting the size of market systems to the local level flies in the face of centerpieces of market logic such as division of labor and comparative advantage. The isolation of different small-scale market systems would therefore have to be enforced in order to prevent their spontaneous agglomeration. Enforcement of this sort implies social choice very different from that of the market, in the form of administration or coercion.

This kind of structural change may be an intriguing prospect, but of itself it is no ecologically rational panacea. A small-scale economic system would, one suspects, suffer many of the coordination problems constitutive of ecological irrationality as long as it is market-based, and characterized by market morality. Moreover, if localized small-scale markets were not tightly controlled, they would always be pressing for their own expansion.

A Moral Persuasion Adjunct

One can imagine a mechanism of moral persuasion grafted onto a market system such that "enlightened" consumer preferences would henceforth prevail. Moral persuasion as a form of social choice is a wholly different kettle of fish from the market, and therefore will be addressed in its own right in chapter 11 below.

A Substantial Role for Government

Even at their *laissez-faire* best, market systems need some form of governmental authority, if only to enforce private property rights and the laws of contract. In contemporary market systems governments do, of course, play a far more extensive role. In the realm of ecological concerns, this more substantial role can be justified by the need for coordination in the control of access to common property resources and in the provision of public goods. Restricting growth, forcing markets to be less responsive to positive feedback, inhibiting interest rates, and regulating market morals are tasks which contemporary governments in market systems do not generally perform. An increased role for government in a market system can take three forms: active intervention by the state, by public enterprise, and by planner sovereignty. These three possibilities will now be discussed in turn.

The state can actively intervene in market systems to correct for their identified failings. This normative position underpins the conception of the state in contemporary welfare economics.[17] To accomplish this task, the state has a number of policy instruments at its disposal: taxation of various sorts, prohibition of activities, subsidies, direct supply of public goods, regulation, and operations on the content and distribution of property rights.

To enhance the ecological rationality of market systems, governments could variously:

improve ambient environmental quality through regulation of pollution,

---

[17] Economists interpret "market failure" as an inefficient allocation of resources. They tend to disagree among themselves as to the actual extent of failure – the more conservative the economist, the less failure she perceives – but agree on this fundamental role for the state. Here I wish to treat failure in a broader, ecological sense.

84   *Evaluation*

per-unit charges on emissions, subsidy of pollution control equipment,
or taxation on the products of polluting processes (see Baumol and
Oates, 1975): such measures involve the direct enhancement of
negative feedback signals;
regulate and coordinate access to common property resources through,
for example, applying permitting schemes for fisheries, limiting extraction of water from underground aquifers, or forbidding high-altitude
supersonic flights;
use the conventional instruments of macroeconomic management to hold
down interest rates or restrict economic growth;[18]
institute a system of severance taxes to discourage the depletion of
non-renewable resources (see Page, 1977);
engage in research and development in ecologically benign technologies,
such as "soft" energy;
legislate and enforce changes in property rights in order to ameliorate
commons problems;
invest directly in ecosystems – for example, by planting trees.

All of these operations should, it seems, be within the capabilities of
governments in contemporary market systems. Yet, despite the occasional clean air act or limited entry fisheries scheme, most governments
make no serious moves in the direction of any of them. Two factors
underlie this inaction. First of all, governments are not the purposive,
integrated actors that the welfare economist – or, for that matter, the
ecologist – would have them be. Both welfare economists and ecologists
have a simplistic view of the state (Young, 1982) which stands in marked
contrast to the subtlety of their analyses of the systems which form the
focus of their study: markets and ecosystems, respectively.

Second, even if they could act in a purposive, integrated fashion,
governments are highly constrained in what they can do within the
confines of a predominantly market system. As Lindblom (1982) points
out, the market is a "prison" which constrains the content of public
policy. Any actions – such as those prompted by ecological concerns –
which damage the profitability of business will immediately be punished
through reduced investment and increased unemployment. Business
confidence is a very fragile thing, and government officials fear to damage
it, knowing they will be held responsible for any resulting unemployment
or recession. Democratic governments fear losing the next election,
authoritarian governments fear a coup or insurrection. The crucial

[18] Unfortunately, it would be hard to do both: low interest rates stimulate capital
investment, the engine of growth.

importance of business confidence in a market system means that business holds a unique, privileged position of access to governments in market systems (Lindblom, 1977: 170–88). This influence need not be in the form of a conspiratorial "power elite" (Mills, 1956); any government official with a modicum of intelligence will be able to predict or react to the market signals of business confidence.

The presence of an active government yields a form of social choice no longer close to a pure market system. Governments come in a variety of stripes, and can incorporate a number of different kinds of social choice mechanism. The lengthy treatment demanded by governmental social choice will be received in chapters 8, 9, and 11 below.

The second possibility for an increased role of government in a market system lies in the introduction of public enterprise. Under public enterprise, corporate executives are replaced by government officials, who are expected to "produce and sell whatever customers will buy, pay for whatever inputs are necessary, avoid losses, cover costs" (Lindblom, 1977: 95–6). The market continues to operate, hence this system is not one of command planning. Examples of partial public enterprise markets are the mixed economies of Western Europe, where government enterprises produce coal, cars, electricity, steel, ships, airline travel, oil, and (until 1971 in Great Britain) some of the best beer to be found anywhere.[19] There currently exists, though, no good example of a fully fledged public enterprise market system.[20] These Western European cases generally have only one public enterprise per industry, which may or may not be a monopoly.

Public enterprise does, in fact, offer little prospect for ameliorating the ecological irrationality of market systems. Replacing private executives with public officials will do little to change collective outcomes if those officials are subject to the same incentive structure and feedback signals as their predecessors. It is evident that public enterprises in Western industrial societies behave in much the same manner as their private sector counterparts (Lindblom, 1977: 112–13), and exhibit no greater sensitivity to environmental values. If public enterprise were to be more rational than private enterprise, then the relevant officials would have to be directed to pursue ecological goals – which in turn would require either command planning or very heavy-handed moral persuasion, mechanisms of social choice far removed from the market.

The third possibility for enhanced governmental participation in

[19] The sale of the Carlisle Brewery was the major act of privatization performed by the 1970–4 Conservative government.

[20] The Yugoslav economy is not a good example, as the internal structure of its enterprises is (at least nominally) democratic.

market-based social choice would replace consumer sovereignty with planner sovereignty. The market would operate as before, but with government planners determining the pattern of final consumer demand (Lindblom, 1977: 98–100). In practical terms, this arrangement means that the government would act as the sole purchaser for all final goods, just as it now purchases tanks for its armed forces, or paperclips for its bureaucracy. Needed too would be a system to distribute these goods to consumers. This distribution system could be conducted by administrators, or through the sale of goods. If the latter option were chosen, the price of goods would rise or fall depending on the amount of each good that the government chose to purchase and sell (Lindblom, 1977: 98–9). Planner sovereignty differs from command planning in that it has no concern with the internal workings of enterprises.

Again, no fully functioning system of planner sovereignty can be found in the real world. Theoretically, though, planners in such a system could control the content of feedback and therefore could direct and coordinate the pattern of demand and production in an ecologically rational fashion. Private cars could be made scarce and expensive, public transport cheap and plentiful. The price of goods intensive in their use of non-renewable resources or energy could be increased, and they could be made less available than renewable resource and labor-intensive goods. Products whose manufacture required polluting technology could be eliminated.

Realistically, such a system of planner sovereignty would face subversion through black markets, which could be eradicated only by repression. Moreover, this kind of system makes heavy demands upon the capacities of government planners – now with greatly enhanced powers – to act in a purposive and ecologically rational manner. To suppose that governments *could* or *would* act thus involves a very naive conception of government. Would planners choose to promote ecological values? Or would they pursue other ends? Should we rely on the good intentions of an elite of planners? Again, the capabilities of governmental social choice cry out for examination.

CONCLUDING OBSERVATIONS

Private enterprise, consumer sovereignty markets may have their good points, but it should be clear by now that ecological rationality is not one of them. Markets of this sort merit unequivocable condemnation for their failure to achieve negative feedback and coordination in their interactions with ecosystems. Further, none of the reforms to market systems explored in this chapter offers any significant promise of improvement. At best, these modifications shift the burden of achieving ecological rationality onto some other form of social choice, such as a governmental hierarchy, moral persuasion, or more anarchistic, communal mechanisms. These alternatives will be explored in the chapters that follow.

8

Administered Systems

Market rhetoric may be sweeping the Anglo-American world, but it has yet to penetrate the Iron Curtain (except in the watered-down form of incentive schemes to promote economic production). Official doctrine in Soviet bloc countries still claims that the market (specifically, capitalism) is the prime *cause* of ecological problems, and certainly not a potential solution. The market is even held culpable for these countries' admitted environmental problems, which are interpreted as legacies of the capitalist past (or, in the special case of the Soviet Union, the tsarist past). The official solution to ecological problems is the same as the official solution to every kind of problem: socialism. For "socialism" here one should read "centrally administered planning."

It therefore remains somewhat paradoxical that there is currently more in the way of administrative planning for environmental goals in the Western world than in the Soviet bloc. For, while the capitalist world has made substantial use of environmentally oriented administrative agencies such as the Alkali and Clean Air Inspectorate in Great Britain and the Environmental Protection Agency in the USA, little beyond lip-service has generally been paid to ecological goals by the governments of the Soviet bloc.

One exception to this generalization may be the German Democratic Republic (East Germany), widely regarded as a bloc model for environmentally sensitive administration.[1] The GDR is the most highly industrialized and densely populated country in Eastern Europe. It has substantial ecological problems, but as the most prosperous Eastern European country it should also possess the economic resources to effect planned solutions to these problems. The government does indeed

---

[1] This discussion of the East German case draws heavily upon Fullenbach (1981).

recognize the severity of water pollution, air pollution, and ground-water depletion. Moreover, in addition to blaming the capitalist past, officials admit that too little attention was paid to environmental goals in the haste of post-1945 economic reconstruction (see Fullenbach, 1981: 15).

The East German government enacted a far-reaching land use law in 1970. The law specifies that natural resource and environmental questions shall be incorporated in the national five-year economic plans, which now have a section entitled "environmental protection." Central administration and coordination of the provisions of the law are in the hands of the Ministry for the Environment and Water Management. Environmental and natural resources policy can therefore include specific directives (such as limits on groundwater use, and one ingenious rule that the water intakes of riparian industry must be located *downstream* from its effluent discharge), the incorporation of environmental objectives in the plans of productive enterprises, and the central planning of specific environmental projects (such as sewage treatment plants).

The actual results of East German environmental administration have been less impressive than the paper product. The scraps of evidence which are available suggest that environmental deterioration was at best stabilized in the 1970s (see Fullenbach, 1981: 19). One major obstacle to improved ecological performance is that the five-year plans still specify productive growth as their overarching goal. If ecological proposals contribute to enhanced production (as may be the case with some water pollution control measures), then they stand a good chance of implementation. If, on the other hand, they obstruct production, they will almost certainly be set aside. This situation is typical of that obtaining throughout the Soviet bloc, where Marxist materialist dreams of productive cornucopia through rational direction of the forces of science and technology die hard.

THE PROMISE OF ADMINISTRATION

One may ascribe the apparent ecological failings of contemporary centrally planned economies to their continued stress on production above all else. Thus, dismay with the workings of market social choice in Western societies can still lead some to a conviction that heavy doses of central control are demanded if the irrationalities of the market – or of decentralized systems more generally – are to be remedied. Central control is normally conceived of in terms of administered social choice: exercise of the authority of an elite through a subordinate bureaucratic structure.[2] Soviet bloc economies are merely extreme manifestations of this kind of social choice.

Disillusion with decentralized social choice can arise for any number of reasons. So, for example, socialists in Western Europe have traditionally recommended an extensive role for government in order to ensure a socially desirable pattern of production and distribution of goods and services. Going back further, Thomas Hobbes stressed the need for a powerful sovereign to exercise authority over society in order to prevent men from destroying one another in their unconstrained interactions (Hobbes, 1946). Moving toward recent decades, theorists of the "post-industrial society" foresee the development of technocracy as the technical complexity of the issues facing society intensifies (Bell, 1973). In this technocratic spirit, Alvin Weinberg (1972) recommends a permanent "priesthood" of nuclear technologists who would exercise control in the interests of securing the safe and effective operation of the centralized energy systems at the heart of the economy of the future.

The ecological realm is not without its socialist planners (Stretton, 1976), its technocrats, and, above all, its Hobbesians. Its best-known Hobbesian is perhaps Garrett Hardin, who advocates "mutual coercion, mutually agreed upon" as the solution to the tragedy of the commons (Hardin, 1968). The ensuing powerful central authority would regulate access to common property resources and limit the freedom to breed.[3] Also in the Hobbesian camp is William Ophuls, who suggests that what is needed is an enlightened, benevolent, but inescapably authoritarian "natural aristocracy" to force society to accept the demands of "ecological scarcity" and to curb the appetites of the ignorant (Ophuls, 1977: 226–7). In comparison to Robert Heilbroner (1980), though, Hardin and Ophuls appear positively cheerful; Heilbroner envisages a comprehensively grim human prospect, which *might* just be alleviated by coercive and totalitarian central control. Even some of those more optimistic souls whose future is a benign and peaceful steady state find it in themselves to "accept the necessity of social controls for the transition period" (Harman, 1976: 143). At a less drastic level, many ecologists see comprehensive environmental central planning as holding out the promise of rational management decisions (see, for example, Odum, 1983: 499–501), even as such planning coexists with decentralized social choice in other functional areas.

---

[2] Note, though, that there are other ways in which a central elite can secure control; see, especially, chapter 11 below.

[3] Later, Hardin becomes yet more pessimistic (or, as he would say, "pejoristic"), in suggesting that all that remains is for the wealthy of the world to organize themselves into "lifeboats" in order to avoid drowning in a sea of human misery (Hardin, 1977).

Ecological Hobbesians vary in the degree of control they would grant to central authority, and in the content of possible checks upon that control. What they share is a faith that such control, if exercised by sufficiently well motivated and knowledgeable people, would indeed alleviate ecological problems, if not set us securely on the road to some stable end-state. The imagery here is of instrumentally rational and potentially effective problem-solving carried out by some integrated and purposive collective entity, such as the state.[4] The backing for this faith now merits examination.

ON BEHALF OF CENTRAL CONTROL

Perhaps the most seductive aspect of central administration is a promise of *coordination*. One would think central planners could cogitate and act in the manner demanded by the non-reducible and collective nature of ecological problems. Coordination would then be effected through the imposition of common purpose upon an administrative structure, an imposition achieved in its turn through the issuing and enforcement of directives to the various parts of the bureaucratic system. Such purpose could quite easily cover control of common property resources (or conversion of those resources to administered public property) and provision of public goods. At first sight, this coordination potential stands in stark contrast to the fragmentary tendencies of decentralized social choice mechanisms such as consumer sovereignty markets and polyarchies.

Coordination across choices in an ecological context must be achieved under conditions of complexity in the environment of a decision system (indeed, complexity is one reason coordination in social choice is necessary; see chapter 4). Administered systems do indeed possess devices for coordination under complexity. Confronted with a complex problem, the appropriate coordinating unit(s) within an administrative structure can divide that problem into its component parts, which may in their turn be divided into subsets (H. Simon, 1981; Alexander, 1964). This subdivision produces a "tree" of sets and subsets into which the initial complex problem is decomposed. Each subset may then be allocated by the coordinating unit(s) to the appropriate operating unit within the administered system. The initial problem, unmanageable in its totality, is

---

[4] This imagery of the state finds its most striking manifestation in the worldview of the welfare economist, where the state is seen as that entity which acts in a rational manner to correct for market failure.

now rendered amenable to piecemeal resolution by those operating units acting in parallel and in concert. Most large human organizations are, of course, structured along exactly these lines. So, for example, a governmental environmental agency might be divided into bureaus dealing with air pollution, water quality, hazardous wastes, and radiation; the air pollution bureau might itself be split into units handling automobile and stationary sources; and so forth.

If coordination rather than chaos is to ensue, any such decomposition of problems and tasks must be an intelligent one. An intelligent decomposition will proceed such that (a) interactions among the elements of a problem allocated to any single subset are rich, and (b) interactions between elements allocated to different subsets are weak (Alexander, 1964: 124). So, for example, an intelligent decomposition of the energy problem might begin according to time horizon. The initial sets would then be defined as immediate crisis, short-term vulnerabilities, and medium-range and long-range supply problems (see H. Simon, 1981: 181). The outcome of problem decomposition followed by solution assimilation would, ideally, be a kind of central planning simultaneously compartmentalized and holistic (see Odum, 1983: 499–501, for some examples in landscape design).

Administered systems also possess two devices for achieving coordination across actors *within* particular choices. It was noted in chapter 7 that solution to the prisoner's dilemma is promoted to the extent that individuals interact on a frequent basis. Specialization, routinization, and a stable division of labor in bureaucracy mean that interactions between particular combinations of individuals are highly repetitive (see Axelrod, 1984: 130). Individuals within any given bureaucratic unit do of course interact on a continuous basis; higher up the organizational ladder, inter-unit activities are brought into concert by managers who themselves interact frequently within a stable group. Task decomposition further promotes this frequency by converting any major decision into a series of smaller decisions.

Administration's second instrument for solving the prisoner's dilemma is more blunt: the players involved can be instructed or coerced to cooperate rather than defect. The dilemma is pre-empted by taking choice away from the players. This procedure is, of course, exactly that which Hobbesians commend to a sovereign power (see above).

Negative feedback can also be organized in an administered system. In contrast to the diffuse and internal feedback controls of ecosystems and markets, feedback in a teleological administrative structure passes through a regnant center. That "controller" need not be the apex of an administrative hierarchy; controllers can be dispersed at appropriate points throughout the structure (for example, in local offices of an agency, each charged with monitoring environmental quality in its area). The building of feedback devices into administrative social choice is essentially an exercise in cybernetic system design.

In further contradistinction to negative feedback in market systems, that in administered social choice need not, one would think, be inhibited by any inherent myopia. Administration is often commended for its ability to engage in long-range planning – presumably grounded in a capability to anticipate long-range problems.

Administration can also lay claim to *robustness*. Hierarchy and administration have sometimes been seen as the supremely rational forms of social organization – those that can cope with a wide range of complex problems. Hence problem circumstances which would devastate the functioning of an administered system should be rare. Empirically, it would indeed seem to be the case that large organizations – especially public organizations – exhibit a degree of robustness verging on immortality. Proponents of technocracy see administration becoming increasingly necessary as the circumstances of social choice become more complex, but there is no reason to suppose that administration is any less desirable in simpler circumstances. Following Max Weber (see Gerth and Mills, 1948), bureaucracy has sometimes been regarded as the hallmark of rationality in social and political organization.[5]

Flexibility, too, might be a feature of administered systems. Administrative reorganization can and does occur. Agencies and bureaus can receive new charges or be relieved of old ones; responsibilities can be re-allocated as necessary; the style and content of intra-organizational interaction can be adjusted; units which have outlived their usefulness can be extinguished; new units can be created in response to the recognition of new problems, or new potential solutions. Note, in this respect, the flurry of activity in the creation of governmental environmental agencies throughout the industrial world around 1970.

If bureaucracy is indeed the supremely rational form of human organization ("rationality" being understood here in terms of a general capacity to solve problems), then *resilience* might be another quality of administered systems. Recall that resilience in an ecological context refers to the capacity of a social choice mechanism to restore or attain the vicinity of a stable anthropogenic subclimax, should disequilibrium conditions exist. Under administered social choice, resilience would take the form of coordinated problem-solving under the auspices of a leadership elite.

---

[5] This view is not particularly popular among contemporary organization theorists.

Who, then, would belong to such an elite? Aside from persons recruited for their administrative talents, the elite would need, one supposes, the established authorities in the scientific field or fields deemed important. Members of Weinberg's "priesthood" or Ophuls's "natural aristocracy" would presumably be selected on the basis of their superior knowledge, whether of nuclear technology or the finer points of ecology. Indeed, Ophuls (1977: 163) explicitly commends a "class of ecological mandarins." In the same way that the "vanguard" in Marxist–Leninist theory is the guardian of the laws of history, this kind of elite could be the guardian of the laws of ecology. The rest of us would presumably defer to that superior knowledge.

The supposition that any kind of knowledge – such as ecological knowledge – is the preserve of an "authoritative" elite rests ultimately on a particular view of the nature of science and knowledge. This view, known variously as positivism or logical empiricism, holds that there exist bodies of scientific truth which can be discovered, verified by experiment (or other empirical means), and added to the stock of human knowledge, to be inviolable thereafter. If scientific knowledge were indeed of this form, and if it were decided that society should be governed according to the laws of a science such as ecology (or economics, or Marxism, or cybernetics, or sociobiology), then a hierarchy subordinate to some scientific priesthood would indeed have a good claim to be the best form of social organization.

That claim would warrant special attention under any disequilibrium demanding resilience. But the claim would hold too under more mundane conditions in the vicinity of a stable operating range in the interactions of human and natural systems. Weinberg's "priesthood" secures stability, as well as attaining it.

A case can be made, then, that all five criteria for ecological rationality in social choice can be met by administered systems. The lessons of experience are more equivocal.

Clearly, administered systems *have* scored some notable successes. Human beings coordinated in bureaucracies can do many things which human beings as isolated actors cannot. It takes a bureaucracy to fight a war successfully, to reach the moon, to build a Concorde airplane, to conduct foreign relations, to organize production in a business firm of any size, to operate a bank, or to provide social services on a large scale. The centrally administered economies of the Soviet bloc have, during their periods of forced industrialization, achieved rates of economic growth never matched by any market system (see Kuznets, 1966) – at least, not until the recent market-based spurts of the newly industrialized countries of East Asia. Similarly, the wartime planning controls instituted during 1939–45 in countries such as Germany, the UK, and the USA were remarkably successful in redirecting national economies to the prosecution of the war effort. At a more modest level, the administered systems (Kapitsa et al., 1977). The lack of ecological motivation within guaranteeing adequate health care for an entire society at comparatively low cost (Maxwell, 1981).

Examples of ecologically successful administered systems are less easily identified. The managers of public lands and regulators of pollution in the USA (and, for that matter, in other Western industrialized countries) have their accomplishments, but also some very severe shortcomings. Ecological values are not generally the prime concern of such agencies. So, for example, the US Forest Service operates under the nebulous "multiple-use" doctrine. The federal Environmental Protection Agency concerns itself only with the direct impacts of pollution on human health. The US Soil Conservation Service in its early years is sometimes seen as an exemplary ecologically successful agency (see, for example, Odum, 1983: 276); however, that agency operated in a largely polyarchical manner. Any such agencies are heavily influenced and constrained too by the polyarchical systems of which they are a part, and by their market context.

Outside of such contexts, some observers are impressed by the ecological potential of the centrally planned system of the Soviet Union (Pryde, 1972; Goldman, 1972: 273). The administered systems of the Soviet bloc are not, though, noted for their environmental sensibilities (see Singleton, 1976; Fullenbach, 1981), despite official lip-service to ecological values and some extravagant claims made on behalf of those systems (Kapitsa et al., 1977). The lack of ecological motivation within the leadership of Soviet bloc systems means, however, that one cannot point to such cases to demonstrate the ecological irrationality of central control *per se*.

This latter point is more widely applicable in that self-consciously ecologically motivated administered social choice has yet to be tried seriously anywhere; therefore, there exists less empirical evidence on which to base an evaluation than in the case of market choice, which can only remain non-teleological. Thus the ecological rationality of administered systems can be assessed only in terms of the likelihood of its realization, rather than *directly* from the lessons of experience. The *indirect* evidence is strong, though – strong enough, as will be argued in the next section, to cast doubt upon the adequacy of administered social choice on all five criteria.

THE SHORTCOMINGS OF ADMINISTERED SYSTEMS

Coordination Difficulties

The leadership of an administered system might be wise and ecologically omniscient, but wisdom and omniscience do not guarantee coordination if leadership lacks effective control over the organizational structure of which it is the apex. Experience suggests that leadership control over the operations of administered systems is problematic, to say the least. Without that control, there is little to guarantee that the parts of a system will act in concert in the manner demanded by non-reducibility and complexity in the ecological circumstances of social choice.

One concerted attempt to redirect several administrative agencies away from ecological values demonstrates how hard it is for leadership to exercise control over bureaucratic structure. In 1982, Ann Gorsuch was installed as Administrator of the US Environmental Protection Agency, and James Watt was appointed Secretary of the Interior. In the eyes of environmentalists, there followed a three-year "reign of terror," which ended only with the forced resignations of the principals in 1983. Gorsuch and Watt directed systematic attempts to replace key officials (both political appointees and career civil servants) adhering to the protective missions of these agencies with individuals more sympathetic to industrial development and economic growth. The hierarchy at Interior proved resistant to Watt's efforts – if with substantial external help (see Culhane, 1984). The effects in EPA were more dramatic; budget cuts and personnel changes combined to cripple the Agency's enforcement of environmental regulations. But the EPA case demonstrates only that it is possible to dismantle an agency and obstruct its mission. This experience sheds no light on whether wholesale redirection of bureaucratic problem-solving energies toward some positive goal is possible.

Studies of bureaucratic organizations in both private (Cyert and March, 1963) and public (Allison, 1971) sectors have generally found a low degree of compliance with leadership intentions in subordinate levels of organizational hierarchy. In the context of the 1962 Cuban missile crisis, Allison found that organizations under the nominal control of the President and Secretary of Defense behaved in a largely autonomous manner, in several cases openly contravening directives from above. Thus the Navy, on being instructed to organize a blockade of Cuba, ran one according to *their* book, thoroughly insensitive to both the spirit and the letter of the instructions of President Kennedy and Secretary of Defense McNamara (Allison, 1971: 128–32). The consequence might easily have been an incident which in turn could have triggered escalation to a nuclear exchange. The Air Force, on being instructed to investigate the possibilities of a limited "surgical strike" against the missile sites in Cuba, prepared to wipe Cuba off the map (Allison, 1971: 123–5). It should be noted that Allison's analysis is developed in the context of crisis decision-making on the part of an active and vigilant leadership group. If leadership control can be frustrated here, one suspects it can be frustrated anywhere. Ecological administration entails a continuous and coordinated stream of actions, as opposed to the dramatic interventions typical of a crisis.

Part of the explanation for this observed lack of compliance of nominal subordinates undoubtedly lies in conflicting interests: the Navy was interested in testing and demonstrating its anti-submarine and blockading capabilities; the Air Force saw Cuba as a thorn in the side of the USA, to be eliminated at the first opportunity. Some analysts even attempt to explain the workings of bureaucracies in terms of fairly naked *individual* self-interest (Niskanen, 1971). Yet self-interest tells only part of the story.

Organizational response to a problem of any complexity requires decomposition of that problem into sets and subsets (see above) and a concomitant division of labor between different units and individuals. Any division of labor, though, makes additional demands upon coordination among the parts of the organization. Coordination can be achieved only through a routinization of tasks, which simplifies the burden of calculation of all the actors within the organization. Routinization leads to the development of standard operating procedures, which constrain the range of responses of units within the organization and that of the organization as a whole. Any command from above which falls outside this range is unlikely to be effective; the only real options which a leadership group has are defined by the routines which the organization is capable of performing (Allison, 1971: 67–100). The freedom of maneuver of leadership is therefore severely constricted.

Instances in which leadership directly orders subordinates to take specific actions – such as Allison's missile crisis case – are, in fact, relatively rare in large organizations. More common are instances in which leadership sets broad guidelines, and leaves implementation to lower levels of the hierarchy. Think, for example, of the social programs of the US federal government; or of its environmental policies, which must be applied to specific and variable cases. Implementation requires interpretation; the nature of hierarchy requires series of interpretations. The members of that hierarchy may have the best will in the world, but a cumulative chain of interpretations will make it highly improbable that the intentions of the leadership will ultimately be translated into practice, and that the actions of different units will be concerted. Any form of administered system is composed of units connected in series, as well as in parallel; such a form of organization is highly brittle, and can be rendered ineffective by failure or misinterpretation at any one point in the chain (Levine, 1972: 139–40). Even if the probability of success at each point is high, in a chain of any length the cumulative probability of compliance can approach zero distressingly easily. In the face of this kind of consideration, some observers express wonder that any US federal programs work at all (Pressman and Wildavsky, 1973: 107).

Confronted with organizational recalcitrance, a leadership group has a number of options. As Levine (1972: 133) notes, the typical bureaucratic response to the unanticipated consequences of directives or rules is more rules; but these additional rules are themselves candidates for interpretation. At best, more rules will add to rigidity and "noise" in the organization, and thus impede feedback. Alternatively, the leadership can attempt to make the content of its directives more detailed and explicit; this approach is likely to saddle that group with an intolerable burden of calculation. It was widely noted that one of Jimmy Carter's major failings as President was his appetite for detail in policy decisions. The central economic planners of the Soviet economy are another group faced with an excessive cognitive load. As the Soviet economy diversified beyond the production of weapons and capital goods, unintended consequences of central directives multiplied, and coordination between different productive enterprises became increasingly difficult to achieve. Despite the introduction of computers and techniques such as input–output analysis, though, the Soviet planners find they still cannot exercise effective control; subordinate units always find loopholes to exploit to their own advantage, or simply respond to the incentive structure as it appears to them.[6] Sporadic decentralization attempts in the Soviet system have generally been aborted as soon as it became clear that decentralization does indeed inhibit central control, moving the system in the direction of market social choice.

The lower levels in an administered system typically have a better (or at least different) conception of the particular problems confronting them than does the apex of the administrative pyramid. Of itself, this factor contributes to a sensitivity to negative feedback signals in the system. However, differences in conception of the problems facing the system mean that subordinate units are easily tempted into giving the illusion of

---

[6] On these and other problems of the Soviet economic system, see Bornstein and Fusfield (1974).

compliance to leadership instructions, while simultaneously going their own way. They may even establish relationships with other units which are informal and transgress boundaries on an organizational chart. As Thayer (1981) notes, most bureaucratic organizations typically possess both a formal hierarchy and an informal set of relationships. It is quite conceivable, then, that coordination may be achieved through a form of collective choice, operating informally, which is largely uncontrolled and cooperative in nature. It is via this kind of informal interaction that cooperative solutions of the prisoner's dilemma may arise. Thayer (1981) suggests that it is through such informal arrangements that useful work may be accomplished in a formally hierarchical context. Lindblom (1977: 67) notes that bureaucracy in practice often resembles a "clumsy market," in which exchange and interaction compensate for the failings of administration. Kelley (1976) finds a degree of pluralist interaction even in environmental policy-making in the Soviet Union.

Any centrally administered system which attempts to exercise *detailed* control over the individual cases in its domain is asking for non-compliance and subversion on the part of its subordinate units. Such subversion may, if Thayer is correct, actually promote coordination and feedback capabilities, and thereby ameliorate the ecological irrationality of administered social choice. One must question, though, the rationality of a form of social choice which relies for its effectiveness on its own subversion.

Subversion of organizational boundaries reflects a fundamental flaw in the way administered systems achieve coordination across choices in the face of complex conditions. The primary means of administrative coordination are problem decomposition and a parallel task allocation (see above). As complexity intensifies, however, decomposition becomes less justifiable. Indeed, highly complex problems may lack *any* justifiable decomposition, for interactions across the boundaries of any conceivable subsets would be too rich. In ecology, everything is indeed connected to everything else. Even what Herbert Simon refers to as "near-decomposable" problems are rare in the ecological realm (H. Simon, 1981: 212–13). If even a thoughtful and sensitive decomposition will fail to turn the trick here, consider the difficulties that ossified and outmoded bureaucratic division of labor will cause.

The consequences of arbitrary decomposition of ecological problems – not least of which is displacement from one medium to another of the sort described in chapter 2 – has recently been recognized by the US Environmental Protection Agency. In response, the EPA has floated the idea of "integrated environmental management," a modelling exercise in which the effects of any single anti-pollution measure are traced across media (see Mosher, 1983). Thus, one could determine the impact upon water quality of an air pollution regulation that restricted sulphur dioxide emissions; the most cost-effective means of achieving a given improvement in human health could then be calculated, taking into account *all* the effects of each anti-pollution proposal under consideration. Any such model would be specific to a particular locality (the EPA has targeted a number of "hot spots").

Integrated environmental management has not met smooth sailing, in part because it would transfer decision-making power from Congress (which currently writes environmental laws) to EPA modellers. So the technique remains unproven, although one suspects it would fall victim to the excessive cognitive burden which plagues the modelling exercises of central economic planners. Integrated environmental management would at best mirror the complexity of the real world of ecological problems, rather than coping with that complexity. In practice, this procedure involves modelling systems from a narrow engineering perspective which ignores ecological relationships. Moreover, it does not fully overcome the hazards of problem decomposition, inasmuch as each exercise must be spatially bounded.

Flawed Feedback

Administered systems are teleological in nature. Therefore effective negative feedback requires the "programing" of appropriate control units with ecological goals; sensitivity on the part of those units to signals from their environment; appropriate interaction and communication within control units; and the crafting of reasoned responses. Failure is endemic in each of these aspects of negative feedback in administered systems. These four aspects will now be examined in turn.

Goal-setting may be fairly easy at upper levels of an administrative hierarchy (but see the discussion of robustness below). However, as noted in the preceding discussion of coordination problems, those norms will not necessarily pervade all levels of organization.

Sensitivity to external signals can also be problematical. Effective negative feedback in an administered system demands a willingness to accept the possibility of unanticipated consequences and error in previous decisions. Indeed, feedback is necessary in man–nature systems precisely because error is *probable*, not merely possible. Individuals within hierarchical organizations are rarely willing, though, to stand aside and let their theories die in their stead. For the reputation and fate of those individuals – and the organizational units of which they are a part – are tied up with the success or failure of the actions identified with them. Failure is not an acceptable mode of learning. Consequently, administered systems often effectively conceal and hence perpetuate their errors. Two of the worst examples of this syndrome are the cover-up of a massive nuclear accident in the Urals in 1958 by the Soviet hierarchy (Zile, 1982: 203), and the less successful denial, first of occurrence and then of severity, of the 1986 near-meltdown of a nuclear reactor at Chernobyl. Note, too, the fear with which the agencies of the US federal government greet any possibility of an external evaluation of their programs (Anderson, 1979: 160). In-house evaluations, though still not always welcomed by program administrators, are far more likely to accentuate the positive, and so pose less of a threat of eventual punishment. The effective evaluation which makes learning from experience possible demands that evaluators look for evidence of failure, not just of success; success is very easy to find if one looks hard enough for it.

Reception of feedback signals is further inhibited by the tendency of organizations to craft a "culture" from their perspectives, routines, imperatives, and operating procedures (Hummel, 1982: 60–98). Embedded in such a culture will be theories of how the world works. Administered systems are quite capable of holding to incomplete or inappropriate theories in the face of manifest failure. The result is that failure can be perpetuated to the point where it becomes truly spectacular, as in the case of the Vietnam war (Levine, 1972: 122–31).

Even assuming that there exist points in an administrative structure receptive to feedback signals, there can be considerable difficulties in transmission and interpretation of those signals *within* the structure.

Individuals within bureaucracies are, among other things, information processors: they receive stimuli (as directives from above, reports from below, or signals from the organization's environment), they cogitate, and they act. The information-processing capacities of individuals are highly limited (Steinbruner, 1974: 58–61). The fact of that limitation is one reason for the organization of human beings into bureaucracies, to carry out tasks of a magnitude beyond the capacity of individuals in isolation: individuals can perform tasks only serially, whereas organizations can do so in parallel.

Members of an organization are people rather than automata; and therefore the way they process information is variable. However, some generalizations can be made. The information-processing capabilities of the human mind are the subject of a considerable literature in cognitive psychology, a review of which is beyond the scope of this study (see Crecine, 1982, for a partial survey). As a first approximation, though, individuals in bureaucratic organizations tend to homogenize their information environments into predictable patterns, so as to facilitate

communication with one another. Individuals behave in highly structured ways, and force new information into familiar categories (Inbar,
1979). Therefore the information circulating within an organization will
often bear a highly imperfect resemblance to the conditions in the
environment of that organization.[7]

Individuals and units within an administered system can, then, remain
in substantial isolation from any knowledge of the real effects of their
actions. Wilensky (1967) notes that large organizations are subject to a
number of "information pathologies"; information is husbanded as a
scarce resource, to be used strategically in intra-organizational struggles.
Moreover, even in the absence of conscious obstruction and deception,
the parts of an administered system – including its apex – can still remain
in considerable ignorance of what is happening as a result of the
organization's actions. Kenneth Boulding notes that hierarchy can be
described in terms of a series of wastepaper baskets in which most useful
information ends up. The more levels in the hierarchy, the more likely it
is that individuals in its upper levels will be operating in a purely
imaginary world (Boulding, 1966: 8). And imaginary worlds are not the
best places to coordinate responses to real-world ecological problems.

Administered systems can expect to encounter some further difficulties
in the crafting of reasoned responses. Coordination in large organizations demands a routinization of tasks which constrains the range of
responses available (see above). Within a bureaucratic organization,
then, a limited range of stock responses is available, and unfamiliar
problems will in all likelihood be re-interpreted to fit that available range.
The fact (noted above) that organizations tend to craft a culture of their
own imposes an additional set of blinkers on the organization's outlook,
and constricts its available set of responses yet further. Under such
circumstances, the normal organizational reaction to a novel situation is
to apply a routine solution with which the organization is comfortable;
only fortuitously will this be remotely adequate. The tendency of the
military to want to "fight the last war" is often noted.

The very size of governmental organizations imposes a further set of
constraints on possible responses. Large organizations tend to think in
terms of large responses to problems, for such responses are a way of
justifying their size and continued existence. So, for example, Exxon is
uninterested in the possibilities of decentralized solar power; one
suspects that solar energy, if ever organized by Exxon, would consist of
Arizona covered with collectors that exported electricity elsewhere, or a

---

[7] For a fictional account that rings true of an organization acting in terms of perceptions
that bear only a tangential connection with reality, see Le Carré (1965).

satellite collecting sunlight and beaming it to earth in the form of microwaves. The Army Corps of Engineers and Bureau of Reclamation are interested in building large dams and irrigation systems, not in creating a more rational spatial distribution of agriculture, housing, and industry. The Department of Energy devotes precious little attention or funds to decentralized, renewable energy sources. Federal responses to the energy crisis have tended to come in the form of giant (if ill-fated) proposals such as the Nixon administration's "Project Independence," or the Carter administration's synthetic fuels program, or subsidized nuclear power. The tendency of large organizations to think in terms of large solutions is shown at its worst in the USSR, where the most likely response to the despoliation of the arid south is the reversal of several north-flowing rivers (Zile, 1982: 206–7). Large-scale projects by their very size threaten to exceed the homeostatic and adaptive capabilities of ecosystems.

Administration as Problem-solving: A Failure in Resilience

As with any form of social choice, shortcomings in negative feedback and coordination might be forgivable if centralized administered systems exhibited a high degree of resilience. It was noted in the previous section that administration's claim to resilience is backed by the notion that bureaucracy is the supremely rational form of social organization. This notion in turn is warranted by a particular conception of science: positivism.

The above discussion of flawed feedback should have already brought to mind considerable doubts concerning the resilience of administration. That discussion suggests that bureaucratic organizations are fairly clumsy devices in an environment of complex, subtle, and ever-changing signals; hardly a recipe for effective problem-solving. To that recognition can now be added the further point that central administration's claim to rationality and resilience is based on a discredited view of the nature of science and scientific inquiry.

The positivist view of science, though not without a place in the popular imagination, is today something of an anachronism. Contemporary philosophers of science recognize that there can be no such thing as *verified* scientific truth; most theories are eventually superseded by more highly corroborated ones. To philosophers such as Popper (1972), the distinguishing feature of scientific problem-solving is open criticism. There are those who also see a function for some incubation and protection of theories from criticism which might prematurely devastate them (Lakatos, 1970; Kuhn, 1962; Laudan, 1977); but at some point any theory simply must open itself up to scrutiny. The rational debate and criticism of ideas at the center of the scientific enterprise can only be inhibited by a scientific priesthood or other guardians of the truth.[8] There is, then, no *scientific* justification for authoritarian central planning (see Magee, 1973: 77–8).

As noted in chapter 3, ecological problems are characterized by high degrees of complexity, variability, and uncertainty. There can exist no universally applicable body of "verified" scientific theory under these (or, for that matter, any other) circumstances. Any theory in the possession of the apex of an administrative structure is therefore going to be a wrong one; to enshrine it in social practice would be a serious mistake. The experience of attempts to turn policy-making over to experts such as systems analysts or program budgeters has been a dismal one; all too often, claims to expertise are based on little more than adherence to some idiosyncratic technique or theory.[9] Note, for example, the failure of "shining city" urban redevelopment, or of attempts to promote equality of opportunity through operations on the educational system in the USA. One of the major causes of failure in public policy implementation is the basing of action on a mistaken theory (Pressman and Wildavsky, 1973: 147–60). Most social theories offer at best a simplified view of a complex world (LaPorte, 1975: 337–8). It is hardly surprising, then, that public policies informed by such theories normally have consequences which surprise their progenitors. The possibility of surprise in itself is not devastating; but, in the absence of a capability to react intelligently to changing signals concerning the outcomes of interventions (feedback), that possibility can engender a severe incapacity in problem-solving.

As Thayer (1981: A14) points out, a theory of organization must always have a theory of knowledge embedded in it. Hierarchical organization presupposes that people in superior positions know more than their subordinates. This presupposition in turn stifles the debate and criticism necessary for effective problem-solving. While hierarchy *may* be adequate for the coordination of routine tasks (but see the discussion of coordination difficulties above), it is a bad problem-solving device. And, in a world in which problem-solving is more important than the coordination of routine – such as a world threatened by severe ecological problems – hierarchy is a bad principle for the organization of society. An effective problem-solving community, the archetype of which is a

---

[8] Such priesthoods flourish in theocracies or societies beholden to some secular ideology; think, for example, of the Catholic Church enforcing geocentricity at the time of Galileo, the Aryan sciences of Nazi Germany, or the crude historical materialism of Stalin's Russia.

[9] In the context of economic policy, Keynes (1936) noted that even "Practical men, who believe themselves to be quite exempt from any intellectual influences, are usually the slaves of some defunct economist."

scientific community, is a community of equals in which good arguments prevail, not the authority of individuals.[10]

Hierarchy and authority obstruct the free and open dialogue in the development and scrutiny of competing tentative solutions to problems which effective problem-solving demands. Superordinates in hierarchical structures typically act so as to maintain a myth of certitude by obstructing free inquiry, repressing opposition, and screening their decisions from open scrutiny. Organizations as diverse as business firms, the Roman Catholic Church, the Soviet Communist Party, and the British Civil Service display these obstructive features.[11] Only if the myth of certitude can be replaced by other grounds for deference, such as those found in some large Japanese organizations, can such obstruction be avoided.

In sum, then, to yield problem-solving in the ecological realm to an administrative structure under the control of an elite on the basis of the supposedly superior capabilities of that form of organization would be a deeply irrational step.

A Fragility of Purpose

Administered systems, as already noted, can exhibit a robustness that verges on immortality. That kind of robustness, though, takes the form of a resistance to change in the existence, formal structure, or entrenched routines of an organization. The type of robustness of interest in this study, in contrast, is a robustness of performance across a wide range of conditions. With respect to any set of external goals (i.e., goals other than organizational survival and growth), administered systems are, in fact, very fragile.

Ecological rationality in administration depends on a clear and shared commitment to ecological goals on the part of the central controlling elite.[12] Devotees of the conspiracy theory of history will find it easy to swallow the possibility of the existence of a cohesive elite; though, historically, ecology has not been among the concerns of conspiratorial elites. The elite envisaged by ecological Hobbesians would, presumably,

---

[10] This recognition does not deny that such communities can be characterized by differentiation according to ability.

[11] The management of large organizations (such as corporations) is, though, sometimes aware of the conditions for effective problem-solving; complex problems are often taken out of the hierarchy and handed for "brainstorming" to non-hierarchical groups only loosely connected to the administrative structure (see Bruner, 1962).

[12] Note the parallels here with the brief discussion of the "programing" of control units with ecological values under "flawed feedback" above.

be somewhat more benign. What are the prospects, then, of a shared and steadfast commitment to ecological values on the part of some such controlling group?

Any ecologically rational administered system needs a considerable (though not perfect) internal consensus covering the values it holds and the relative weight they should receive. However, the normal condition of politics in both centralized and polyarchical political systems is conflict over values. So equality confronts economic efficiency in poly-archies; "reds" battle "experts" in the Chinese Communist Party; the Catholic Church suffers the occasional schism and continuous (if muted) internal dissent. Any teleological system requires consensus on values if it is to produce some good consistently and effectively. Under administered social choice, the difficulty is compounded because that commitment must be a *continuous* one. The administrative structure cannot waver in its commitment, for it is only that commitment which can keep the system on course.

History suggests that political elites are often less than unbending in their shared commitment to any goals more novel than defence of the *status quo* in the distribution of economic spoils. This generalization applies to incoming radical governments of the right or left in poly-archical societies, where the demands of economic and political marketplaces can all too easily douse the flames of radicalism.[13] The radical anti-environmentalism of the executive branch of the US federal govern-ment in the early 1980s fared little better. Such cases are not, though, a particularly good test of elite steadfastness, given their inbuilt inhibitors of administrative social choice. A better test of that steadfastness may be offered in cases where elites come to power committed to *truly* novel values; such cases are normally (if not necessarily) revolutionary. Post-revolution experience does, however, often parallel post-election experi-ence in the extent of the eventual de-radicalization or re-orientation of incoming elites (Tucker, 1967). The histories of the French, Mexican, and Russian revolutions are instructive in this respect. Even the Chinese revolution succumbed, despite the best efforts of Mao himself in his declining years. One shining exception to this rule is Nazi Germany; with a few lapses, such as an accommodation with capitalism, it did indeed stick to the principles of its founders.

One could find exceptions other than Nazi Germany, but the general point is that elite groups are highly vulnerable to redirection away from any radical or novel set of values. Authoritarian regimes, unless they take some very draconian measures, may be still more vulnerable than

---

[13] See discussions of markets and polyarchy in chapters 7 and 9, respectively.

incoming administrations in polyarchies in this respect. Dictators live in constant fear of coup or insurrection, and so must always attend to day-to-day survival concerns; the ensuing myopia, worse than that of any conceivable market system, further inhibits effective negative feedback. Governments in polyarchies have to worry only about the next election – or, if they are unlucky, about the next vote of confidence in parliament.[14] Moreover, members of polyarchical governments need fear only the reduced salaries of the opposition benches, as opposed to the firing squad.

One must conclude, then, that the steadfastness of any hypothetical ecological elite, while not entirely inconceivable, is a highly unreliable basket in which to place all one's eggs.

A Rigidity of Structure

Administration's lack of performance robustness might be compensated for by flexibility in organizational structure. While organizational redesign in administered social choice can and does occur, it is an extremely cumbersome process. It was stressed in the above discussion of coordination difficulties that bureaucratic organizations are highly resistant to redirection from above. Organizational routines and structures are not readily amended to fit novel conditions. While there may be a degree of "drift" in routines and structure, that drift is likely to take the form of ossification as an organization ages. To take one example, the US Soil Conservation Service succumbed to administrative arthritis after its initial ecological successes in the 1930s (see Odum, 1983: 276). From an ecological perspective, a slightly more positive note is struck by the resistance of the US Department of the Interior and the Environmental Protection Agency to redirection from above in the early 1980s.

CONCLUSION

The summary judgment on administered systems is that, despite a very superficial promise, they fail on each of the five criteria constituting ecological rationality. That failure may, perhaps, be debatable in the cases of robustness and flexibility; indeed, any shortcoming on robustness may be indicative of flexibility, and vice versa. Failure is most acute in negative feedback, coordination, and resilience. The net result is

---

[14] The latter kind of concern is most central in systems where multi-party coalition government is the norm, such as Italy, Belgium, and the Netherlands.

that administered systems are adept at dealing with only two types of problems.

First of all, it is clear that such organizations deal well with familiar, routine, and static problems; indeed, they may be the optimal form of collective choice under such conditions. Successful examples would include bureaucracies that collect taxes, administer agricultural subsidies, distribute welfare payments, organize an army in peacetime, and construct interstate highways.

Second, administered systems can mobilize resources quickly and effectively in the pursuit of unambiguous and tangible goals such as defeating an enemy in wartime, initiating a crash program of capital investment, reaching the moon, or building a supersonic transport.

Highly structured organizations are at a loss, though, when it comes to dealing with high degrees of uncertainty, variability, and complexity – circumstances that are, of course, ubiquitous in the ecological realm. Thus, new side-effects of chemicals are continually being discovered (and disputed); ecosystems start to behave in unanticipated ways (for example, as eutrophication of a lake sets in); unsuspected interactions between familiar chemicals emitted into the environment come to present a major hazard; chronic ecological damage suddenly becomes apparent (for example, as occurred in the case of acid rain, or DDT); or infrequent events such as a major nuclear power plant accident or an oilspill occur. Moreover, what is good practice in Arkansas may be ecologically ruinous in Arizona or Alaska. Uncertainty concerning the future consequences of present actions, such as the generation and disposal of nuclear wastes, the discharge of persistent toxins into the environment, or the depletion of a key natural resource, compound the difficulties. Signals from natural systems are subtle and ever-changing; the rigidity of a bureaucratic organization is a poor way to cope with them. The stock solution will rarely be the right one.[15]

Administered hierarchies are, then, an inappropriate form of social choice under the conditions which constitute the ecological circumstances of social choice. Indeed, the fate of hierarchical command under circumstances of uncertainty, non-reducibility, complexity, and

---

[15] Some good examples of the agencies of the US federal government acting in generally inappropriate ways are afforded by the experience of natural resource management in Alaska. Thus, prospective oil and gas lease sales are evaluated by the Bureau of Land Management on a tract-by-tract or sale-by-sale basis, with little thought given to the overall coherence of resource development in previously undeveloped regions. Environmental effects of development are assessed in terms of their direct impact upon local human populations rather than upon the wilderness which constitutes much of Alaska. See Dryzek (1983a).

variability is often a spontaneous (if incomplete and resisted) dissolution into more consultative and collegial forms of interaction (La Porte, 1975: 354–5; see also Thayer, 1981). Such dissolution is greeted by enthusiasts of "matrix organization" in administrative reform. Matrix organization circumvents hierarchy by creating impermanent groupings of individuals around particular tasks (the implications for ecological rationality will be addressed in chapter 14).

If hierarchy is dissolving in real-world organizations attempting to cope with such circumstances, one should hesitate before considering the wholesale increase in hierarchy that ecological Hobbesians commend. Any such attempt would, in all probability, undergo a similar dissolution. This recognition is of no small comfort, for it banishes the specter of "technofascist" ecological engineering, which Gorz (1980) fears on the grounds of its potential effectiveness in resolving ecological problems.

The ecological irrationality of administered systems stems from two clusters of factors, one associated with the very existence of a leadership elite, and the other concomitant with the exercise of control through bureaucratic, tightly coupled organizations. Any purported rationality of elite control is rooted in an improbable steadfastness in pursuit of ecological values and, more important, in an error in conception of the nature of human problem-solving. Bureaucratic organization compounds the irrationality of elite control due to the improbability of effective coordination, the limited feedback sensitivity and problem-solving competence of bureaucratic organizations, and a resistance to structural change.

The prospects for ecologically rational centralized, administered forms of social choice are bleak indeed. Such systems can perform a limited number of tasks quite well. But, as Lindblom (1977: 76–89) recognizes, authority systems tend to have "strong thumbs, no fingers." Ecological rationality demands green and nimble fingers: where may such fingers be found, if not in market or administered systems? The quest continues in the next chapter.

9

Polyarchy

Polyarchy is the form of social choice which defines the countries often thought of as "democratic" – most of them West European or Anglo-American. Polyarchy is not, however, the universal form of social choice in any society; and rarely, if ever, is it even the dominant form. Rather, polyarchy co-exists with other forms of social choice, especially markets and administered systems (administered systems can be component units in polyarchies, just as they can in markets). The mix of market, administration, and polyarchy can vary considerably in its proportions. So, for example, whereas Britain may be categorized as a polyarchy, its government is far more centralized, and therefore contains larger doses of administered social choice, than does the USA. Governments of the Soviet bloc, though normally classified as centrally administered systems, can in fact exhibit some aspects of polyarchical interaction. In this chapter, though, my intention is to explore the ecological strengths and weaknesses of polyarchy *qua* polyarchy, rather than of permutations that include polyarchy.[1]

Polyarchy, is more easily recognized than defined. Instead of offering any succinct definition, prominent polyarchists such as Dahl (1971: 3) and Lindblom (1977: 132–3) list a number of characteristics which must be more or less satisfied by a system if it is to be called a polyarchy. This list includes a familiar set of individual liberties, free and meaningful elections, and – crucially – the freedom to join or establish organizations, associations which in turn should be capable of exercising political influence. For present purposes, the key aspect of polyarchical mechanisms is that collective choices are the outcomes of interactions between relatively large numbers of actors, none of which is capable of exercising

---

[1] There are those who argue that polyarchy in the real world is at best a facade, at worst a deception, and that real control in such systems rests in the hands of an economic elite (see, for example, Parenti, 1983; Domhoff, 1978). If the reader accepts such arguments, she is referred to the discussion of elite control in chapter 8.

anything remotely approaching authoritative control over the system. Interaction should proceed in the context of relatively free exchange of argument, information, and influence, such that choices are arrived at through "mutual adjustment" between partisans of different interests (Lindblom, 1965). The interplay of actors takes place against the background of a set of rules – such as a constitution – but many of its elements will be informal. The prohibition on authoritative control means that no actor (such as an economic elite) can be dominant, but neither is absolute equality among actors and interests required.

POLYARCHY IN A NUCLEAR WASTELAND

Unlike the two forms of social choice discussed so far, polyarchy has a nuclear waste problem. Market forces, if left alone, would not permit as unprofitable an activity as nuclear power production.[2] In contrast to the market, administered systems have promoted both nuclear power and nuclear weapons, and have acquiesced in the generation of nuclear wastes. However, as long as these wastes were under control of the administrative hierarchy in question, they remained a mere technical concern. Thus, in the United Kingdom, British Nuclear Fuels reprocesses radioactive wastes (at its Windscale–Sellafield plant in Cumbria), and disposes of radioactive effluent through a pipeline leading into the Irish Sea. The weapons programs of the US military have been generating high-level wastes since 1945. For several decades, disposal of these wastes was not regarded as a particularly interesting issue by anybody – let alone the military. This activity was generally treated as a national security matter, and so exempt from the scrutiny received by similar kinds of non-military actions (see Goodin, 1982: 228–9). One supposes that the Soviet-administered system does not agonize unduly about the disposal of its nuclear wastes.

In contrast to the market, which would not produce nuclear wastes, and administration, which causes their generation but cares little about their disposal, polyarchy suffers prolonged anguish over the issue. Thus, while Britain blithely reprocesses spent fuel from commercial reactors, the USA ceased such operations in 1977 under direction from President

[2] The economics of nuclear power is shrouded in controversy. Its defenders deny that other forms of electricity production are cheaper, and blame excessive regulation and political interference for spiralling costs. Yet the fact remains that nowhere has a nuclear power plant been installed and operated without substantial public subsidy, either in the form of insurance, tax breaks, research and development funding, government sanction of inflated rates to captive electricity consumers, or public construction and operation.

Carter, sensitive to the proliferation of weapons-grade nuclear material associated with reprocessing. In Britain, environmentalists upset with waste generation and disposal can only resort to extra-legal protest and sabotage; Greenpeace has attempted to block the Windscale pipeline and to physically obstruct ocean dumping of low-level wastes. In the USA Greenpeace and other anti-nuclear groups can lobby, receive a sympathetic hearing, and have some of their concerns responded to in public policy. But even in the USA, as long as nuclear wastes remained a matter for the military hierarchy, their problematic features received little publicity. Only when commercial reactors began generating similar kinds of wastes did substantial political fallout occur. Today, the whole issue of high-level nuclear waste is a subject of protracted, lively, and seemingly irreconcilable debate and dispute in the American polyarchy.

High-level radioactive wastes do indeed present some intractable problems. Some are solid; some are liquid; most are long-lived (with half-lives of thousands of years); all are highly dangerous (see Walker, 1983, for the relevant technical details). And nobody knows how to dispose of them safely, for "safety" here must mean near-total isolation from the biosphere.

High-level wastes have been accumulating in the USA for over four decades, but no permanent disposal has yet been made. Instead, temporary storage is the order of the day; spent fuel sits inside canisters in pools of water at reactor sites, and military liquid wastes are stored in tanks in South Carolina, Idaho, and Washington State. Options for permanent disposal include polar ice sheets, the seabed, outer space, and deep burial in rock formations. No terrestrial method is demonstrably foolproof over the thousands of years the waste would have to remain isolated. Migration into the biosphere is a possibility even for deep-buried waste, due to the cumulative impact of faulting, the corrosiveness of the wastes themselves, and groundwater flow. The considerable uncertainties involved cannot be ameliorated through incremental trial and error, for the cost of error is high, and it could be a matter of centuries before meaningful results were generated.

The record of the American polyarchy in dealing with the nuclear waste issue is, on the face of it, unimpressive. Once all nuclear issues were the preserve of a tight and exclusive "iron triangle," which, as part of its pro-nuclear ways, managed to keep the waste question out of the limelight. This triangle was composed of the Joint Committee on Atomic Energy of the US Congress, the federal Atomic Energy Commission (AEC), and the nuclear industry. However, following the 1974 division of the AEC into the Nuclear Regulatory Commission and the Energy Research and Development Administration (later incorporated into the

Department of Energy) and the 1977 dissolution of the Joint Committee, the iron triangle's power wilted (see Woodhouse, 1983, for further details). Thus, "corporatist" control over policy gave way to more truly polyarchical social choice, as more actors and interests were admitted to the policy arena. (The difference between corporatism and polyarchy will be explained below.)

The actors and interests involved in the radioactive waste issue are numerous and varied in their outlook. They include federal agencies such as the Department of Energy, Nuclear Regulatory Commission, and Environmental Protection Agency; the Interior, Energy and Commerce, and Science and Technology Committees of the US House of Representatives; the Energy and Natural Resources, Environment and Public Works, and Governmental Affairs Committees of the US Senate; various state and local governments; corporations; nuclear industry associations such as the Atomic Industrial Forum and Committee for Energy Awareness; citizens' groups; and anti-nuclear groups such as the Union of Concerned Scientists, Friends of the Earth, and Clamshell Alliance. Pro- and anti-nuclear forces both have their friends in government. So the Department of Energy and the House Committee on Science and Technology tend to be pro-nuclear; the House Interior Committee leans toward the anti-nuclear position (see Woodhouse, 1983).

Many aspects of the waste disposal issue have received thorough scrutiny in the prolonged and lively debates engaged in by these actors. Collective choices can respond to the range of arguments advanced and political influences exercised. However, jurisdictional rivalry, sharp political conflict, and the widespread exercise of veto power over policy proposals have led to impasse on the high-level radioactive waste issue.

The current solution favoured by the federal excecutive branch and the nuclear industry is a central underground repository to receive all the USA's high-level wastes. Environmentalists have fought this option because *any* resolution of the wastes issue is one less argument against the idea of nuclear power. Environmentalists are encouraged here because several states have passed laws prohibiting the construction of new nuclear power stations until a permanent solution to the waste problem is found. With one or two exceptions, no state or local government wants the repository in *its* backyard – even if it recognizes that the dump must go somewhere. Readily available veto power in the American polyarchy means that the repository may not be located in *anyone's* backyard.

In 1984 the US Department of Energy announced a short-list of three sites for the repository: the Hanford Nuclear Reservation in Washington State, Yucca Mountain in Nevada, and Deaf Smith

County in Texas. These sites were chosen on political as much as geological grounds. The Hanford area's economic mainstay is already nuclear, and Nevada residents should be thoroughly inured to the vicissitudes of the nuclear age, given the state's history of weapons testing. But despite the attention to local feelings in site selection, none of the three sites faces smooth sailing. The governor of Nevada announced he would fight the Yucca Mountain site; the governor of Washington expressed extreme reservations about Hanford; and the governor of Texas proclaimed that "before the people of Deaf Smith County will glow in the dark, sparks will fly."[3] State governors can veto a site proposal, but any veto cast can be overridden by a vote of both Houses of Congress. According to the Department of Energy timetable, the first waste should be buried in 1998. But such timetables have come and gone in the past, and there is little reason to expect resolution of this issue in time for 1998.

The history of the radioactive waste issue is not, it seems, a good advertisement for the problem-solving capabilities of polyarchy. Equally clearly, though, polyarchy out-performs both markets and administration on this issue. No pure market system (even if it *did* produce nuclear power) would possess any plausible means of coping with high-level wastes. Administered systems have succeeded only in denying the problem's existence.

There may be a more general lesson here: that the incapacity of both markets and administered systems means that *their more visible and intractable ecological problems are displaced to polyarchy.* Markets hand their problems to government (see chapter 7), and "government" means polyarchy rather than administration for truly severe and complex problems like nuclear wastes. Even in the Soviet Union, the few environmental issues which do manage to evoke a high degree of concern may be subject to "polyarchical" resolution. For example, the most prominent ecological issue in the USSR in the 1960s and 1970s was perhaps the condition of the unique ecosystem of Lake Baikal in Siberia. The lake was threatened by industrial development (especially paper mills) in its drainage basin. "Grass-roots" organizations of scientists and concerned citizens moved to protect the lake, and had some success in pushing Soviet government policy in that direction.

Sometimes, then, polyarchy may function as the social choice mechanism of last resort (or at least, next resort) for the ecological failings of market and administration. This function suggests that polyarchy may be better able to handle ecological problems than these two prominent

---

[3] See "Department of Energy Picks High-level Nuclear Waste Sites," *Not Man Apart* (February 1985: 15).

alternatives. In this chapter I will attempt to show that, despite some severe shortcomings, polyarchy's ecological rationality is indeed greater than that of market and administration.

THE ATTRACTIONS OF POLYARCHY

Just as Western economists admire markets, polyarchy tends to be the Western political scientist's favourite brand of social choice.[4] It is sometimes noted that pluralism (a variant of polyarchy) is, in effect, the prevailing ideology of American political science (for example, Winner, 1975: 71). Undeniably, the ideal type has a considerable number of virtues, which real-world polyarchies can be expected to approximate in varying degrees.

Polyarchy can lay claim to ecological rationality primarily on the grounds of negative feedback and resilience – though polyarchy also possesses a unique form of coordination. If validated, these grounds would constitute a strong case for polyarchy, even in the absence of much in the way of robustness and flexibility.

Consider first of all negative feedback. Feedback in ecosystems and market systems is diffuse and internal; that in administered systems is centralized. Polyarchy contains both diffuse and centralized feedback devices, and therefore falls at an intermediate point on a spectrum from teleological to non-teleological systems.

The non-teleological aspect of negative feedback under polyarchy resides in the ability of individuals to organize and exert pressure within the political system. Should any category of people consider its interests slighted, or feel that those interests merit stronger promotion, then that category can organize an interest group. One would expect intensity of political effort to depend on extent of concern: people choking on pollution or living close to a proposed high-level nuclear waste repository are likely to expend greater energy than those suffering a minor loss of amenity.

Categories and groups can, then, seek to exercise political influence and hence bend collective decisions toward the values they hold. The greater the abuse of the natural systems upon which a category of people (for example, coastal fishermen, or residents in the vicinity of a nuclear plant) depends, the more powerful the negative feedback signals, and the greater the likelihood that collective choice will be directed toward rectification of the ecological abuse in question.

Contemporary polyarchies also possess a number of environmentalist groups, generating a continuous stream of negative feedback on a variety

---

[4] It should be noted here that a majority of Western political scientists resides in the USA.

of ecological issues. One of the more successful groupings of this type was the American "conservation movement" of the 1900s, which reached its zenith in the presidency of Theodore Roosevelt. Since the late 1960s environmental groups have mushroomed in number and size. Today in the USA one finds the Sierra Club, Natural Resources Defense Council, National Wildlife Federation, Audubon Society, Wilderness Society, and numerous local and regional groups; in the UK there are the Council for the Protection of Rural England, the Royal Society for Nature Conservation, and the Royal Society for the Protection of Birds; in Germany there is a "Green" political movement which collected 5.6 percent of the vote in a 1982 national election; and, internationally, there are Friends of the Earth and Greenpeace. These groups struggle politically for various aspects of environmental quality and ecological integrity.[5] The content of environmental legislation, such as the National Environmental Policy Act, Clean Air Act Amendments, Water Pollution Control Act, Endangered Species Act, and Alaska National Interest Lands Conservation Act in the USA, the Clean Air Act and Rivers Pollution Act in Great Britain, and the Basic Law for Pollution Control in Japan, is in large measure a consequence of the efforts of those groups. Many of these pieces of legislation would have been unthinkable before the 1960s. Environmentalism as a political movement reached a peak around 1970, at which time politicans throughout the polyarchical world clambered eagerly into the environmental bandwagon. Those heady days have long since passed, but environmental groups today often have sufficient weight to block any retreat on the commitments of that period.

To further illustrate non-teleological negative feedback under polyarchy, consider the experience of US environmental policy in the early 1980s. It was widely perceived – and not just by environmentalists – that the Environmental Protection Agency under Ann Gorsuch and the Department of the Interior under James Watt would promote federal government complicity in wholesale environmental abuse. The result was a massive increase in the membership of environmental groups, extensive lobbying and publicity campaigns by these groups, and an outbreak of Watt–Gorsuch-bashing in Congress and the media. In 1983 both Gorsuch and Watt resigned under this pressure, and the Reagan administration took the opportunity to return federal environmental policy to the vicinity of its pre-1981 range. Even prior to their resignations, though, Watt and Gorsuch had been effectively fought to a standstill in Congress and the Courts.

---

[5] Few struggle for either in their entirety. For a brief survey of the very different perspectives of American environmental groups, see Dryzek (1983a: 33–4).

Diffuse feedback under polyarchy operates in large measure through the exercise of political influence, but one aspect of that influence is the power of argument. Argument can be couched in terms broader than the narrow self-interest of consumers and producers in the marketplace; the citizen of a democracy is a more virtuous character than the consumer in a market.[6] Moreover, the quality of information is high. Variation in the information flow occurs as actors enter and withdraw from the political fray as they see fit, their entrances bringing new perspectives, new values, and new arguments. The information pertinent to any decision is sifted, highlighted, and organized into arguments by the stakeholders involved. In consequence, the burden of cogitation suffered at high levels in an istered system is avoided. Diffuse and internal negative feedback under polyarchy imposes little strain upon the information-processing capacities of individuals in formal governmental decision-making positions.

Centralized feedback in administered systems is inhibited by hierarchy and bureaucracy (see chapter 8). Centralized feedback in polyarchy should suffer no such fate. For a well-functioning polyarchy constitutes the nearest real-world approximation to a Popperian "open society" (Popper, 1966), outside of course the scientific communities on which the open society is modelled. For Popperians, the open society constitutes the best available form of problem-solving device, for it is only under its conditions that tentative solutions to social problems can be freely stated, criticized, and tested. According to this view, public policy should proceed in terms of "piecemeal social engineering" – actions informed by (highly imperfect) social science knowledge should be taken, evaluated, and modified as seems appropriate. This form of trial and error is, in effect, a thinking person's incrementalism, with great attention paid to feedback from the results of actions. It is crucial for the operation of the open society that there be no common purpose imposed on the system; everyone is allowed to pursue his own ends (except for ends antithetical to the idea of the open society), and to propose, criticize, and evaluate public actions from any viewpoint. This freedom of perspective reflects the Popperian view that the best judges of collective actions are the people affected by those actions, rather than experts acting on their behalf (see James, 1980).[7]

---

[6] The field of public choice eschews this distinction by treating governmental decision structures as just another avenue through which individuals may pursue their preferences. See Mueller (1979) for a survey of the field.

[7] Centralized feedback shades into decentralized feedback here. Feedback is likely to remain centralized only insofar as a common definition of the problem at hand is accepted, and a clear allocation of formal authority to a single actor exists.

The "open society" facilitates negative feedback through free evaluation of proposals and actions. This process of conjecture and refutation in problem-solving also constitutes polyarchy's claim to *resilience*. If any profound disequilibrium in human systems as they interact with ecosystems is recognized, individuals within or outside government can and will investigate and develop possible solutions (think, for example, of the number of proposals floated for dealing with high-level nuclear wastes). Doubtless, many hypotheses will turn out to be blind alleys, but the open society can discard or modify unsuccessful policies. Successful policies will be retained, such that society can move incrementally toward the amelioration of ecological problems. A polyarchy might, then, be able to correct any disequilibrium conditions and move incrementally to more stable conditions in the man–nature system (cf. Johnson, 1979). Thus McClosky, after noting the ecological failings of "totalitarian" states, claims that "In democracies, there has already been a good deal of swift, decisive action in ecological matters . . . because they allow free discussion, dissemination of information, agitation for reform, and protests when reform is not forthcoming" (McClosky, 1983: 157).

Polyarchies have a distinctive way of achieving *coordination* across different decisions in collective choice, despite a lack of any central coordinating device. At first sight, only incoherence across decisions is apparent. So a decision to allow exploration for oil and gas in a frontier region may coincide with a decision to set aside land that would be required for a pipeline system as a national park; a decision to subsidize the operations of a car manufacturer made as a result of pressure from management and unions may be contemporaneous with a decision made elsewhere to discourage car ownership and subsidize public transport on the basis of pressure from city residents, environmentalists, and transportation agencies. (See Wenner, 1976: 120–3 for some further examples.) This kind of process is generally referred to as "disjointed incrementalism." Social choices will be incremental changes from the status quo due to the large number of conflicting pressures, disjointed because there is no centralized coordination among choices in different parts of the system.

Defenders of polyarchy (for example, Wildavsky, 1966, 1979) claim that administered coordination in a governmental system of any size is an impossibility,[8] and that one of the virtues of disjointed incremental collective choice is decentralized coordination in the form of "mutual adjustment" between different actors and decisions. So Lindblom (1965:

---

[8] The analysis of administered systems in chapter 8 would appear to back this contention.

10) believes that the claim of interactive systems to rationality – the "intelligence of democracy" – rests on their achievement of coordination through this kind of reciprocal adaptation. Therefore initially inconsistent actions and choices will, given time, accommodate one another. A degree of coordination forever beyond the reach of the cogitation of any administrative elite can therefore be achieved with comparative ease (see Lindblom and Cohen, 1979).

What, though, of the content of these coordinated outcomes at the system level? It is quite clear that polyarchy is the quintessentially *politically* rational form of social choice (see chapter 5); polyarchy possesses a quasi-invisible hand producing politically rational collective outcomes as a byproduct of individual actors pursuing their own interests. Those outcomes consist of compromises whose content reflects the relative weight of the political resources – votes, campaign contributions, number of activists, moral acceptability, expertise, commitment, bribes, arguments, legal powers, and the like – which organized interests and governmental actors (such as agencies) bring to bear. An analogy is sometimes drawn between the outcome of this kind of process and the position or movement a solid body will exhibit as a resultant of the forces upon it.

This type of outcome minimizes the sum of the discontent of the actors with a stake in a decision, weighted by the political resources at the disposal of each actor. Politically rational processes simultaneously "deliver the goods" to the interests with a stake in each decision, and minimize the alienation of key actors from the political system in question (see chapter 5). Political rationality thus conceived lies outside the scope of administered systems, for any calculated search for decisions on the part of a single leadership group will be insensitive to the weight of organized interests, the subtlety in their pattern of organization, and any incipient interests.[9]

Political rationality need not coincide with ecological rationality. Coincidence will occur, though, if negative feedback signals are of the form discussed above – that is, emanating from human concern about the productive, protective, and waste-assimilative qualities of ecosystems. Further, polyarchy's potential for coordinating access to common property resources and the supply of public goods is clearly greater than that of the market. This potential is partly a matter of polyarchy's capability to produce authoritative decisions at the system level. More important,

[9] The failure of planning, programming, budgeting systems (PPBS) in the US federal government may be interpreted in terms of the alienating consequences of attempting to impose administered collective choice on political, interactive processes (see Schick, 1973; Wildavsky, 1979).

though, is that, *over time,* users of a common property resource (or potential contributors to a public good) can accommodate their actions through "mutual adjustment." Solution to the prisoner's dilemma is facilitated to the extent that these actors constitute a small group interacting on a recurrent basis. The net result can be movement toward agreed-upon controls over the use of the resource.

Robustness in polyarchical social choice is not prima facie obvious. However, there is no cause for immediate concern; for, as noted in chapter 4, robustness is more easily noticed in the presence of a limiting factor, rather than in the absence of any such factor. So any argument on behalf of robustness can most readily take the form of a refutation of suggestions of fragility. Fragility will indeed be suggested (if left unconfirmed) below, so a contemplation of robustness is best postponed to that point.

Flexibility in a polyarchy exists inasmuch as conjectures concerning structural change are readily generated, if less easily tested.

Even setting aside robustness and flexibility, though, negative feedback, coordination, and resilience would together constitute strong backing for the ecological rationality of polyarchy. Moreover, the empirical evidence clearly gives some credence to those claims. While no polyarchy has wholeheartedly embraced ecological values, progress toward rectifying environmental abuses has been more pronounced in polyarchies than in any other kind of extant and widespread form of social choice.

THE FAILINGS OF EXISTING POLYARCHIES

The free and equal access of the polyarchical ideal is not always approximated in practice. *Effective* influence in policy decisions is frequently restricted to a small number of interests, with some dire consequences for negative feedback and resilience. So, in the UK, the "umbrella" organizations of the Trades Union Congress and the Confederation of British Industry can have a very cosy and exclusive relationship with central government, which in turn pales in comparison with the closeness of government and business in Japan. Moreover, even if a real-world polyarchy does possess a large number of organized interests, there is no guarantee of free interplay of interests and arguments. A nominally polyarchical government can in practice fall prey to segmentation into a number of issue areas – agriculture, defense, transport, energy, and so forth – within which a small number of powerful interests (if not a single interest) effectively monopolize political influence, often in a largely conspiratorial relationship with the responsible government agency (and, in the USA with interested legislators). The US Environmental Protection Agency under Ann Gorsuch offered a particularly blatant example of such a relationship, as polluting industry was invited into the inner circles of agency decision-making. The "iron triangle" which long controlled nuclear energy policy in the USA has already been described. A form of social choice in which such patterns of political access prevail is more accurately termed "corporatism," or "corporate pluralism," than polyarchy (Lowi, 1969; Schmitter and Lehmbruch, 1979).

One consequence of any corporatist tendencies is that the content of feedback, and hence policy outcomes, are systematically skewed in the direction of a small number of powerful interests. In the USA and Japan, these interests will normally be large corporations; in Europe, trades unions may also secure a place at the corporatist table.[10] The poor, ethnic minorities, and others lacking financial or organizational resources will in general be excluded from effective influence in the policy process.[11] Environmental groups, too, have a hard time getting a foot in the door, especially in systems with corporatist tendencies as strong as those in the UK (see Enloe, 1975: 264–316) and Japan (Kelley, Stunkel, and Wescott, 1976: 180–96). Corporatism may explain why nuclear waste disposal is not a particularly salient issue in British politics, at least in comparison with the American case.

Some interests may be so powerful that they need not even go through the motions of pluralistic or corporatist interaction. Crenson (1971) interprets the "un-politics" of air pollution in US cities prior to the late 1960s in this light; air pollution, though clearly damaging to human health, was excluded from the political agenda through unspoken deference to the political power of the industrial polluters.

The net consequence of any corporatist (or corporate pluralist) tendencies in a nominal polyarchy is distortion to the point of elimination of diffuse and internal negative feedback. The feedback burden is therefore shifted to more centralized devices. Unfortunately, corporatism's obstruction of free and open debate can cripple centralized feedback too. Resilience will suffer a similar fate. Corporatist social

---

[10] Strictly speaking, corporatism as a form of organization – especially as proclaimed and practiced in fascist systems such as Mussolini's Italy – requires a tripartite partnership between business, workers' organizations, and the state.

[11] This recognition should not be taken as implying that the poor will always come away empty-handed. To minimize their discontent and alienation from the system, government will direct real (Piven and Cloward, 1972) or symbolic (Edelman, 1977) rewards in their direction.

choice resembles administration more than polyarchy, and so merits a condemnation similar to that received by administered systems in chapter 8.

Corporatism is not the only potential villain, though. Even if relatively free and unconstrained political debate should take place, feedback under polyarchy may still be skewed insofar as the actors seeking political influence and having their concerns responded to by the content of collective choices are self-interested in their motivation. Often, the interest is one that can be expressed in tangible financial terms – the interest of the residents of the Richland area in Washington State in the jobs that a nuclear waste dump would provide, of businesses in profits and government contracts, of labor unions in collective bargaining laws, of the poor in welfare payments, of ethnic minorities in access to better-paying jobs. Even when the concern is not readily translated into monetary terms – as, for example, when a community group expresses opposition to the existence of a toxic waste dump in its vicinity – the interest normally remains a *special* one, confined to that group alone. Signals pertaining to *general* values, such as the protective value of natural areas, the productive potential of dispersed common property resources, or even economic efficiency, fall by the wayside inasmuch as they have no special interest attached to them.[12] These general values are diffuse and may be in the interest of large numbers of people, but they may be in nobody's *special* interest. General interests find little reflection in a system of political rationality.[13]

The domination of special interests inhibits centralized feedback mechanisms by debasing the language of politics, ensuring that debate remains at the level of interests rather than underlying and common values. The currency of polyarchy consists of tangible benefits to specific groups; so political argument is couched in those terms. Occasionally, tribute may be paid to values such as the "national interest" or "common good," but such appeals are often little more than rhetorical flourishes, or, at worst, a cover for self-interest – as in "what is good for General Motors is good for America."

At a more subtle linguistic level, the stress of polyarchy on rewards to special interests and hence human want-satisfaction can affect the form of argument used even by environmentally minded interest groups. Thus, the Snake River should not be dammed because of the enjoyment that

---

[12] For an examination of this phenomenon in the context of resource management in the USA, see Dryzek (1983a: 40–1).

[13] A partial exception to this rule can sometimes be noted when polyarchies go to war, and individual or group interests are willingly sacrificed for the war effort. It is noteworthy that President Carter, exasperated by stalemate in energy policy, called for "the moral equivalent of war."

identifiable people get from its present state; a third London airport should not be built at Stansted or Wing because of the damage it will cause to the quality of local life; the dumping of nuclear wastes in the North Atlantic threatens the livelihood of Spanish fishermen; offshore oil development should not take place in California because the scenery will be marred. In couching arguments in these terms, environmentalists are implicitly accepting that environmental values can be traded off against other human wants – hydroelectric power, the convenience of London travelers, nuclear energy, or oil supplies. Environmental values become commodity values, and centralized feedback on the condition of man–nature systems will not be generated. At the bottom of this slippery slope is cost–benefit analysis, which offers a single metric to which want-satisfaction values can be reduced. And cost–benefit analysis is rooted in economic rationality. Under ecological rationality, the whole has to be more than the sum of its special-interest parts. Those interested in the promotion of ecological values should take great care in the language they use (see Tribe, 1974; Bookchin, 1980, *passim*).

Feedback in contemporary polyarchies is further inhibited by a tendency to focus on immediate possibilities and consequences. This phenomenon may be observed in action in locations such as Washington and Westminster, where legislators, their aides, bureaucrats, lobbyists, analysts, and others are continually struggling to meet deadlines. Business is always pressing. People caught up in this kind of system lack any opportunity to reflect upon long-term consequences of actions or the overall pattern of decision. The pressing and the urgent often appear, in retrospect, also to define the class of the ephemeral and trivial. At most, individuals look ahead to the next election; at worst, to that evening's television news. The stress throughout is on short-term feedback at the expense of long-term performance.

The result is that polyarchies can be insensitive to a number of important ecological signals.[14] Falling into this category are the "sleeper effects" of the generation of nuclear wastes (note how long it took the issue to reach the political agenda) and discharge of cumulative poisons into the environment; "threshhold" or "tipping" effects, as arise when, for example, a chemical may be harmless at 60 parts per million, but lethal at 80; and the cumulative effects of decisions such as those which contribute to the buildup of carbon dioxide in the atmosphere. Reactive polyarchical social choice lacks the capacity to cope with these kinds of questions (see Goodin and Waldner, 1979: 2–11).

---

[14] Technically, such signals might be referred to as "feedforward" rather than "feedback."

Any preponderance of special interests in a polyarchy has a number of further consequences potentially devastating to both coordination and negative feedback. The "game" of polyarchical interaction based on self-interest is a distributive "zero-sum" one, in which differential rewards are received by identifiable interests, and one actor's gain means another actor's loss. Polyarchy constitutes an excellent mechanism for the distribution of the costs and benefits of governmental activity. However, the rules that polyarchies develop to facilitate the distributive game are a positive encumbrance when coordination in pursuit of some general value (such as the quality of a common property resource, or the provision of a public good) is demanded. In such instances, the game is "positive sum" – like the prisoner's dilemma described in chapter 4. The widespread availability of veto power over policy proposals – especially in the USA, where it is reinforced by the constitutional system of checks and balances – while functional in a distributive game, can cause paralysis when it comes to the resolution of non-reducible problems demanding coordination in system response. One of the foremost friends of polyarchy recognizes that this kind of incapacity may "put us on the road to catastrophe" (Lindblom, 1977: 347).

The distributive substance of the polyarchical game also introduces considerable positive feedback into social choice: contemporary polyarchies share the addiction of market systems to economic growth. All real-world polyarchies co-exist with market systems. Political conflict is central to the smooth operation of polyarchies, but such conflict must remain muted if the stability of a system is to be guaranteed. Disadvantaged groups must not press too strongly against maldistribution and injustice emanating from the market system. Economic growth is a "political solvent" (Bell, 1974: 43) which enables the discontent of the poor to be bought off at no cost to the rich. Growth lubricates distributive conflict resolution, and enables polyarchies to avoid awkward questions of distributive justice. If growth ceases or becomes negative, then polyarchy is endangered; the great depression saw a number of polyarchies fall victim to authoritarianism. Just as in market systems, the positive feedback of growth is an encumbrance which any negative feedback devices must overcome.

The fact that virtually all polyarchies to date have operated against the background of market systems gives empirical support to those who argue that polyarchy *requires* a market (for example, Hayek, 1944; Friedman and Friedman, 1962). If this contention holds, the prospects for ecologically rational polyarchical social choice are bleak indeed. Not only will polyarchy share the market's positive feedback, but any claims to robustness, resilience, and flexibility will have to fall by the wayside.

For if polyarchy requires a market, it is fragile rather than robust. Moreover, the scope for any structural change involving the parameters of market social choice would be severely limited, calling into question the flexibility of polyarchy. The general problem-solving capacities which constitute the resilience of polyarchy will suffer too.

It was noted in chapter 7 that markets can "imprison" governmental social choice, and polyarchy is not immune. Thus, a class of extremely important collective decisions will simply not be made on a polyarchical basis if a market system is present: "Pluralism at most operates only in an unimprisoned zone of policy making" (Lindblom, 1982: 335). Polyarchy will normally yield to the imperatives of the market, if not always to the interests of large corporations. In defense of polyarchy, though, it has yet to be proven that a market system constitutes a *necessary* condition for polyarchy, the best efforts of Hayek and Friedman notwithstanding (see Lindblom, 1977: 162–8).

In sum, then, it is apparent that existing polyarchies fail (to greater or lesser degrees) to measure up to the ideal. At their corporatist worst, polyarchies degenerate into caricatures of the ideal, with some dire consequences for ecological rationality.

THE FAILINGS OF IDEAL POLYARCHY

It is, in a sense, illegitimate to talk of an "ideal-type" polyarchy, for, as noted at the outset of this chapter, the whole concept of polyarchy is rooted in a largely inductive analysis of some real-world systems. Nevertheless, the shortcomings of existing polyarchies outlined in the previous section stem from a number of identifiable factors: a small number of excessively strong interests, the content of the interests represented, and the market context in which those systems operate. In this section I will attempt to assess the ecological rationality or otherwise of polyarchy purged of these factors. Any such "ideal polyarchy" would feature unrestricted access for a large number of interest groups representing all conceivable social categories and values, with no power of veto, and no market system to co-exist with. A number of factors stand in the way of any claims of the ideal polyarchy to ecological rationality; these obstacles will now be considered in turn.

Feedback and Special Interests

Even the ideal polyarchy will contain *some* actors pursuing their own self-interest. Consider, then, what happens when a polyarchy is presented with an ecological choice opportunity in the form of (say) an oil embargo, a nuclear accident such as Three Mile Island, or a Love Canal or Times Beach. Crisis will shock the system into a flurry of activity, as various actors are presented with an opportunity to present their interpretations and pet solutions, and perhaps to make material gains. Corporations will stress the need to protect business, and will react moderately; regulators will search for regulations to adopt and apply; labor unions will press for the protection of existing jobs; environmentalists may argue for radical action; and so forth. Assuming, in our ideal polyarchy, that there exist no interests powerful enough to exercise veto power, the outcome of these pressures in a polyarchy will be a granting of rewards or protections to all the interests involved. So, in the USA, the 1973 energy crisis led to regulated prices for northern consumers and car manufacturers, largely untaxed windfall profits for oil companies, tax breaks for energy-conserving building modifications for environmentalists, a relaxation of pollution regulations for industry generally, and extended responsibilities for government agencies. The net result was exactly as one would expect in a polyarchy: actions to protect (and, if possible, enhance) the pre-crisis position of the actors and interests involved. The major oil companies, for example, came out of the 1970s crises with added strength and profitability. Environmental values may get some concessions – think, for example, of the legislation enacted throughout the polyarchical world during the "environmental crisis" period around 1970 – but only in proportion to the political resources of actors pressing for them. Crises are effectively treated by the system as occasions for coping with threats to business as usual, rather than as opportunities for any more fundamental change.

The net result of this kind of process is negative feedback of a sort, but pertaining only to stability and homeostasis in a *politically* rational sense. From an ecological perspective, what is happening is *positive* feedback (i.e., deviation amplification). If in the context of problem A a special interest group has its position strengthened, that group will be in an improved position to exert influence and have its concerns responded to in subsequent problem B. Over time, the dynamics of any such process entrench established interests (including those with little concern for ecological values), obstruct coordination, and inhibit high-quality negative feedback.

Resilience in the Garbage Can

Resilience under polyarchy takes the form of a collective problem-solving capacity. Even at their best, though, polyarchies are somewhat cumbersome problem-solving devices. The disjointed and interactive style means that polyarchies can react but slowly to any problems confronting them; compromise takes time, as a large number of "mutual adjustments" may be needed.

The ponderous nature of polyarchies is a severe shortcoming under dynamic and variable circumstances. Aside from the attendant inhibition of negative feedback, any fundamental disequilibrium conditions in man–nature systems are not easily corrected. To be effective, interactive social choice demands a great deal of continuity in the nature of problems confronting a system, a continuity that is lacking in the ecological sphere (see chapter 3). As the pace of change quickens, the qualities of interactive problem-solving disintegrate (see Dror, 1964).

The problem-solving capacities of polyarchy suffer for another reason. Cohen, March, and Olsen (1972) describe the characteristic form of social choice in a loosely coupled system – of which an open polyarchy is an archetype – in terms of a "garbage can" metaphor. The garbage can itself represents an occasion on which the system must produce a decision – perhaps a sudden crisis, perhaps a situation determined by a legislative calendar. The ensuing "choice opportunity" enables various actors to throw problem interpretations and candidate solutions into the garbage can. Note, for example, how environmentalists use the nuclear waste disposal garbage can as an opportunity to advance their opposition to the commercial use of nuclear power. Actors, problems, and solutions move in largely independent trajectories; they mix in an arbitrary way in each garbage can, and so choice opportunities are not easily described in problem-solving terms. Solutions look for problems, rather than vice versa. The collective choices that emerge do so in a largely unpredictable manner; therefore their content is less easily described than in the analogous case of "invisible hand" mechanisms in markets. Certainly, there is no ecologically rational "invisible hand" plucking choices from each garbage can.

Coordination, Complexity, and Common Purpose

The disjointed style of response to public problems under a pattern of special interest representation (noted above) means that certain aspects of problems – those pertaining to specific interests – will receive considerable attention. That attention does not of itself guarantee coordination across the problem as a whole, though. A failure to achieve coordination here can stem in large measure from two old friends, public goods and common property resources, which plague any form of decentralized social choice, not just markets. Resolution of the prisoner's dilemma in solving these problems becomes more difficult as the number of actors involved increases (see chapter 7). The provision of public goods is problematical under our ideal polyarchy simply because of the large number of self-interested actors present. Indeed, the closer one gets to free and open interplay of interests, the harder it becomes to secure the supply of public goods. Every group wants the benefits of collective political choices, but nobody wishes to pay for them. Clean air may be in the interests of a labor union's members, but not at the expense of danger to their jobs. Common property resources – for example, Western grazing land under the nominal authority of the US Bureau of Land Management – can expect abuse in a polyarchy, if actors seek access without responsibility, and pursue their own group interest rather than any conception of the general good. The effective management of common property resources in a polyarchy is further inhibited by conflicting ideas about what constitutes the best use of a resource or ecosystem, not all of them sensitive to its ecological integrity (cf. Crowe, 1969); consider, for example, the number of different uses to which a forest can be put.

These difficulties notwithstanding, it is clear that polyarchy can cope with common property and public goods problems far better than markets can. Under polyarchy, dynamic processes of mutual adjustment and compromise can facilitate cautiously cooperative behavior among actors, and thereby an amelioration of the prisoner's dilemma. Such behavior is out of the question in markets, where competition reigns supreme.

Special interests, public goods, and common property tell only part of the story, though, when it comes to coordination difficulties in a polyarchy facing ecological problems. Dynamism and complexity tell the rest of the story; and these two aspects of the ecological circumstances of social choice can be absolutely devastating to coordination under polyarchy.

Any polyarchy is characterized by division into subsystems – defined by issues, interests, values, or formal authority – each of which contains "watchdogs" (Lindblom, 1959: 85) to promote and protect the relevant interests and values. In our ideal polyarchy, all interests and values – from the profits of General Dynamics to the working conditions of steelworkers, economic efficiency, or environmental quality – would have a subsystem taking care of them. If that polyarchy as a whole is to adapt successfully to changing conditions and problems, then those subsystems must be able to react to change *independently* of one another (Alexander, 1964: 41; see also Ashby, 1960). In a simple, moderately dynamic context this condition may be readily met.

As complexity and dynamism increase, then so does the interpenetration of these subsystems; at some point, *any* decomposition into subsystems will become inappropriate.[15] Choices reached in any one subsystem will inevitably affect the values and interests of concern in other subsystems. Lindblom's "watchdogs" will be barking up each other's trees. The affected subsystems will in turn react with more and different choices of their own, demanding in their turn further corrections at other points in the system. *Given time,* the system might converge on some equilibrium; but a dynamic environment does not offer the luxury of time. Hence difficulties and anomalies accumulate, multiply, and demand corrective action. The continued operation of each subsystem intensifies complexity (Ruggie, 1975: 136–7) by adding to the elements and interactions to be coped with by other subsystems. The obvious escape is simultaneous multiple correction; but such action is not in the repertoire of polyarchy.[16] Together, then, complexity and rapid change are lethal to interactive social choice (see also Winner, 1975: 73).

Consider, for example, the energy problems of the 1970s. In the USA, the overall energy problem was "decomposed" according to the orientations of different interests, ranging from New England oil consumers to the Pentagon. The elements into which the problem was decomposed therefore included short-term dependence on Middle Eastern oil; the shock to the economy of high prices; distributional inequities in the effects of shortages and price increases; national security; excessive consumption; excessive reliance on depletable fossil fuels; and insufficient long-term supply. Unfortunately, little coordination between different aspects of the attack resulted. Actions taken within each subsystem had negative ramifications across other subsystems; think, for example, of the effects of price regulation on the search for new oil supplies or the level of total energy use. The affected subsystems were thereby forced to introduce measures such as reduced environmental controls on energy exploration and development, subsidies for conservation efforts, and contingency plans to invade Saudi Arabia. Fragmented and incoherent responses were often at odds with one another. Instead of convergence on effective solutions, the product was self-perpetuating chaos.

Confronted with a devastating combination of complexity and dynamism in ecological problems, the last remaining possibility for the salvation of coordination – and, for that matter, effective problem-

---

[15] See the discussion of problem decomposition and complexity in chapter 8.
[16] Some observers prescribe authoritarian control on the grounds that it does possess such a capacity (see chapter 8).

solving – in polyarchical social choice would be a consensus on the primacy of ecological values (including the need for coordination) across all subsystems. Such a consensus might be achieved either if (a) ecologically minded interests were to become sufficiently powerful to dwarf all others, or (b) if the power of reason were to hold absolute sway in political interaction, free from any exercise of political power. The prospect of (a) is unlikely, to say the least, and even if it could be obtained it would press upon those ecological interests the requirements of omniscience which were discussed and dismissed as unattainable in chapter 8. Alternative (b), by purging the claims of *any* selfish interests from the system and specifying that only "general" values can be represented, takes us so far from ordinary conceptions of polyarchy that it is better characterized as a totally different form of social choice.[17] One of the defining features of polyarchy is an absence of common purpose; indeed, this absence is seen as one of its prime virtues by advocates of the open society such as Popper (1966). Liberals tend to see value conflict as both inevitable (Unger, 1975: 76–7) and desirable (Williams, 1979). Polyarchy cannot be driven by any single goal. Therefore the fate of ecological values in a polyarchy is to be severely compromised by other values. Under polyarchy, the whole is always the sum of its parts, only a few of which are ever likely to be ecological in inspiration.

A liberal devotion to the multiplicity of human purposes is, under most circumstances, highly laudable. The paradox is that, unless the members of a polyarchy accept a common ecological purpose, then all other human purposes are endangered. Polyarchies are prone to disasters other than Popper's *bête noire* of sweeping vision leading to authoritarianism.

Robustness and Flexibility Revisited

Given coordination and feedback failures, questions of robustness and flexibility again become somewhat moot. For what it is worth, though, robustness would appear to be problematical. No *actual* polyarchy has flourished without material prosperity and a relatively free consumer sovereignty market. However, it is hard to make a case that polyarchy *must* depend on such conditions. Flexibility is something of an open question. While there can be no centrally directed institutional redesign in a polyarchy, it was noted above that conjectures and arguments concerning structural change are readily generated. The problem is that the positive feedback resulting from a pattern of special interest representation will also inhibit structural change.

---

[17] This line of reasoning will be explored further in chapter 15.

CONCLUSION

Once again, the summary conclusion is a negative one. Polyarchies do possess negative feedback devices, but they are far better at responding to signals from General Motors or the *Daily Mirror* than to messages from ecosystems. Systems governed by political rationality lack the capacity to coordinate responses to complex problems – beyond, of course, coordination in the continuing achievement of political rationality. Societies dominated by polyarchy are quite capable of joining those societies and civilizations which have perished rather than respond to ecological signals.

A sense of despondency is lightened, though, by some of the promise that polyarchy holds (even if much of that promise failed to stand up to close scrutiny). In particular, the capabilities of the "open society" ideal in social choice have yet to be exhausted; the open society will re-emerge in part III of this study.

Polyarchy does not, then, merit the unequivocal condemnation received by markets and administered systems; polyarchy's negative feedback and coordination devices are superior to those of both markets and administration. Moreover, polyarchy can lay a stronger claim to resilience.

10

Law

Deforestation is one of the more acute ecological problems facing the tropical world. Now, loss of forest cover is of itself no reason for ecological alarm bells. After all, the forests which once covered much of the temperate world have now mostly disappeared, especially in Europe and eastern North America. In some cases, stable (and perhaps even ecologically rational) man-created agro-ecosystems have replaced the primeval forest.

Tropical rain forests are different. These forests are typically underlain by thin and poor soil. Once the trees are cut and the remaining vegetation burned, the cleared area can support crops for a season or two, and perhaps grazing for a few years thereafter. But the soil's nutrients are soon depleted, and the only way they can be restored is through regeneration of the forest ecosystem. And indeed, a cycle of slash–burn– grow crops–leave fallow–regenerate has long supported "swidden" agriculture in the tropical moist forest biome. Unfortunately, present-day population and commercial pressures do not allow the long fallow period needed for regeneration. In many cases, the resulting over-exploitation leaves in its wake a "red desert," where once was lush forest. Watersheds are left unprotected, which can mean soil dehydration, topsoil loss, flooding in low-lying areas, reduced potable water supplies, and siltation. The more extreme possibilities include genetic depletion and climatological change (see Myers, 1980). Thus the productive, protective, and waste- assimilative capacities of ecosystems can all be decimated by defores- tation.

Tropical deforestation presents, among other things, a *legal* problem, for one of its causes is an inappropriate definition and enforcement of rights and rules. And if one source of the problem is legal, so might be part of the solution; specifically, a redefinition of rights and rules may be required. Consider, in this context, the differences between the legal systems that govern land and trees in temperate and tropical countries.

Costa Rica is illustrative of the tropical case.[1] Deforestation is especially severe there; if present trends continue, forest cover will completely disappear within a matter of years in this once-99 percent-forested country. The severity of the problem is widely recognized within the country, and the government is in principle committed to doing something about it. In this context, Costa Rica is widely hailed as one of the more progressive tropical countries; its admirers point to the large areas of forest protected in national parks.

Despite these good intentions, amelioration of Costa Rica's deforestation problem is hampered by the entrenched legal order. Historically, forested lands were treated as unrestricted common property – an eminently sensible arrangement when population was small and forest seemingly unlimited. Common law in Costa Rica does recognize private property rights, but only if the land is both occupied and "improved" by cutting trees or planting crops. "Possessionary" rights come with a single year's occupancy; land occupied and worked for ten years gives legal title to the occupier; but, if *another* individual purchases the improvements, that party receives title immediately.

This structure of legal rights applies to all kinds of land – common, private, and public – including land notionally set aside in government forest reserves and national parks. A change in landownership can be effected through "squatting," which is an organized business. Squatters can make a quick profit by occupying land and selling "improvements." Fearing invasion of their land, established private landowners carry out clearances of their own, in order to pre-empt the squatters. Some large landowners even organize squatter "armies" to protect the fringes of their property.

Operating in this Costa Rica case is a rather perverse incentive structure. The mechanism for establishing private title is clearance, which, given the poor soil of tropical forest ecosystems, involves destruction of the land's resource potential. Any landowner engaging in protective practices such as allowing the forest to grow back risks losing title.

This odd incentive structure is not without countervailing laws. There are thousands of pages of environmental law on Costa Rica's statute books which would seem to thoroughly protect forest ecosystems. For example, the 1969 Forestry Law requires that all land clearance must receive a permit in advance from the Forest Service. Unfortunately, the formal content of these laws is irrelevant, for they are never enforced.

[1] This discussion of Costa Rica draws heavily on Dryzek and Glenn (1987).

Given that there is no effective penalty for transgression, and that Costa Rica's administrative and legal systems lack the resources for effective enforcement, there is wholesale non-compliance with the provisions of environmental laws. In contrast, the common law traditions which sanction squatting and clearance are widely accepted in Costa Rican society.

Specification of a legal order which would improve on this sorry state of affairs is easy enough. Costa Rica should simply adopt the kinds of rights, rules, and enforcement capabilities governing land and forest use in the industrial world, where all forests are either publicly or privately owned, and no landowner would then need fear invasion. Laws stipulate what landowners can and cannot do. In some cases trees may be protected; sometimes there may be restrictions on clearing steep slopes in a watershed. In most industrial countries the landowner can reap the rewards of good husbandry (though public forests in the USA present some exceptions to this rule, and spillover effects are always a possibility).

The Costa Rica case suggests that law does not change speedily in response to changing circumstances. An inappropriate legal order seems to have outlived its usefulness, but there is little sign of effective reform in the offing. One might argue that Costa Rica's incapacity here is part of its status as a Third World country, short on governmental resources (though by Third World standards Costa Rica is fairly prosperous). Therefore the failure of its legal order in the case of deforestation is no proof of the ecological irrationality of law *per se*.

Several different morals could be drawn from this story. It could be argued that what Costa Rica needs is a fully articulated legal system, rather than its current poorly developed "law of the jungle." Less dramatically, one might argue that all that Costa Rica requires is a reform of its existing structure of rights and rules. More pessimistically, one could give up on law as a means of achieving ecological rationality in social choice, at least in Third World countries. The ambiguity that persists suggests that a thorough scrutiny of the capabilities of legal social choice is in order.

LEGAL SOCIAL CHOICE

Law may be conceived of as a mechanism for making social choices through the development and application of a set of consistent, authoritative, and general rules. Those rules can specify required, permissible, and prohibited behavior; penalties for transgressing any limits; and liabilities for the effects of actions which impinge upon the interests of others. A rule is therefore a guide or standard for action which a category of people is expected to adhere to under specified circumstances (see Young, 1979: 2–3). Any rule normally involves respecting a *right* – that is, an entitlement arising from the occupation of a given role (Young, 1979: 104). Therefore any specific right has a corresponding rule, and vice versa.

Beyond a set of behavioral prescriptions, or primary rules, a legal system requires three additional features. First of all, it must have some notion of authority, whether sovereign or otherwise, to determine the content of specific rights and rules. Authority can be based on common law tradition, judges themselves, or legislatures. Second, some means for changing the content of rules is necessary. Third, some mechanism – usually a court system and the sanctions at its disposal – is needed if rules are to be applied and enforced in specific cases. Hart (1961) refers to these three characteristics as the "secondary rules" of recognition, change, and adjudication which any well-developed legal system must possess. For Hart, any system lacking secondary rules cannot be called "law."

Hart's distinction will be followed here. In order to distinguish law from other forms of social choice based solely on a widespread acceptance of norms (or other types of decentralized enforcement), the focus of this chapter will be upon law in terms of rules promulgated and enforced by some binding authority:[2] that is, legal systems containing both primary and secondary rules. Systems containing primary rules only will be discussed in chapters 12 and 16 below.

Though we occasionally speak of a "government of laws, not men," it is hard to imagine a society in which law is the sole mechanism of social choice. Empirically, law appears as a specialized subsystem which is an adjunct to other forms of social choice; a legal system can clarify and enforce the decisions of a polyarchy, secure the rights and rules necessary for the effective operation of market, or ensure compliance with the directives of administrative structures. A school of "legal positivists" (for example, Hart, 1958, 1961) would see the function of a legal system as simply applying rules made by some other authority (such as a legislature or constitutional convention); from this viewpoint, law of itself is not really a social choice mechanism of standing equivalent to those discussed in the preceding three chapters.

Nonetheless, law is worthy of treatment as a form of social choice in its own right. The fact that a legal system rarely stands alone is not sufficient

---

[2] Some legal anthropologists (e.g., Roberts, 1979) might be unhappy with this distinction, on the grounds that conflict resolution through adherence to rules can occur in systems without formal institutions.

to disqualify it; one might say the same of polyarchies and markets. Moreover, the admonitions of legal positivists to the contrary notwithstanding, courts can and do make social choices that are largely independent of legislative or constitutional directives. First of all, there exists in any well-developed legal system a body of judge-made law, composed of accumulated precedents. In a changing world, the problems of the present are rarely the same as those of the past. Judges therefore have discretion in their choice and interpretations of precedents and rules when faced with a novel case – establishing, in turn, further precedent (see Rosenblum, 1974: 121). Second, even in those cases where courts are called upon to apply a specific piece of legislation to a given case, that legislation is often unclear in both the details of its prescriptions and its very intent. In practice, then, courts have considerable latitude in interpreting legislative intent (see MacCallum, 1968). Vagueness and ambiguity in the content of legislation are, in fact, functional (or "politically rational") in a polyarchy because they are conducive to the establishment of a coalition large enough to pass a bill. In natural resource and environmental legislation in the USA, therefore, one finds meaningless platitudes such as "optimum yield" for ocean fisheries (1976 Fisheries Conservation and Management Act), "proper balance" between competing interests in a resource (1978 Outer Continental Shelf Lands Act Amendments), or "multiple use" as the guiding principle for the management of national forests. None of these terms has any intrinsic analytical or moral content; so the courts find themselves inescapably in the business of moral judgment (see Ward, 1972: 229), as they must decide exactly what "proper balance" should be struck between conflicting interests and values.

Law is, then, a social choice mechanism of considerable scope, especially in a society in which polyarchy is the dominant governmental form. Indeed, until the 1960s the courts carried the burden of environmental social choice in most polyarchies; the question was conceived of in terms of balancing the rights of polluters and victims. The role of the legal system in social choice is currently most extensive in the USA, where virtually all questions of domestic policy end up sooner or later in court as a result of actions brought by individuals or groups slighted by the content of policy proposals or actions. The situation is not quite as extreme in polyarchies with a lesser tradition of judicial independence, or with disciplined parties which have no need to shrink from explicitness in the content of legislation.[3] American pre-eminence in judicial social

---

[3] Coalition-building may still be necessary within superficially disciplined parties if agreement on legislation is to be secured.

choice is reflected in its number of lawyers per capita, and the extent to which they pervade other arenas of social choice such as administration and polyarchical interaction. The USA approximates a "government of lawyers, not men."

THE VIRTUES OF LAW

The prevalence of legal systems may easily be explained. The existence of a body of law acts so as to generate convergent expectations which in turn lubricate and regulate social and economic interactions by defining the bounds of acceptable behavior. Over time, a "legal order" will develop consisting of detailed rights, rules, and regulations. Moreover, law is a relatively bloodless device for resolving disputes between competing claims and interests. As a form of social control, law is less arbitrary and repressive, hence more likely to achieve ready compliance, than authoritarian mechanisms; especially if a legal system is regarded as neutral and fair by the people subject to it. Law's claim to rationality is, first and foremost, its ability to settle disputes when all other methods fail (Diesing, 1962: 146).

This account does not, though, explain why law should be called upon to make wide-ranging collective choices, as opposed to deciding cases involving small numbers of identifiable actors. Arguably, the number of policy choices made by courts in the USA attests to the weaknesses and incompetence of the American polyarchy. On the other hand, law may have some positive attractions which account for its extensive role. Those attractions lead some observers to recommend that government relinquish its role in making environmental policy to the courts (for example, Sax, 1971).

From the vantage point of ecological rationality in social choice, law's two primary attractions are the coordination and robustness it embodies. But negative feedback and even a degree of resilience may be achieved in a legal order.

Coordination

Coordination is a direct consequence of the spontaneous tendency of legal systems to elaborate an order of rights and rules. With the passage of time, the content of those rights and rules – including rules of precedence when principles conflict – will become more fully articulated. Each resolved dispute adds to the legal order, and has the potential for clarifying the structure of that order.

The net result of this process is a consistency of legal rights and behavioral prescriptions standing in marked contrast to the disjointedness of polyarchy. Under polyarchy, collective choices are subject to continual revision and dilution as organized interests fight over the adoption and implementation of policies. In contrast, a directive emanating from a court has the "force of law" behind it. A strong legal system can better enforce compliance, which in its turn will promote coordination, *if* decisions and directives are based upon (or contribute to) a consistent set of rules. A fully articulated legal order might therefore be able to achieve a unified and precise notion of the common good which could inform decisions across the whole range of ecological problems.

Compliance achieved in this manner bears a superficial resemblance to central coordination in administered systems. The key difference is that a directive with the force of law can secure compliance more readily, and at much lower decision cost, than an administrative edict. Compliance here is not just a question of fear of sanction; if a law is clear and precise, most actors will comply simply because expectations converge around that law. A recognition of the power of law in this context is central to the idea of "juridical democracy" proposed by Lowi (1969), under which a legislature would issue clear *and detailed* statutory instructions to administrative agencies. However, the power of law in achieving compliance need not be restricted to securing more effective administration. For law can also be an alternative to administration. Thus, in the context of environmental policy, Dales (1968: 55) suggests that "a change in the law is likely to be a far simpler and cheaper procedure than the establishment of a public pollution-control agency."

Changes in the law of this sort can, if necessary, cover rights of access to common property resources, and rules concerning contributions to the provision of public goods. So, for example, "use and enjoyment" rights of access to a commons (such as a tropical forest) could be restricted as pressure upon the resource intensifies; or a rule that land despoiled by open-cast mining be returned to its original state could be promulgated; or disposal of toxic wastes in watercourses can be prohibited; or landholders might be subjected to a rule that a specified proportion of their land be kept in "protective" rather than "productive" condition. Hence a legal system possesses ready solutions to the common property and public goods coordination problems that plague markets and polyarchy. In the language of the prisoner's dilemma introduced in chapter 4, law specifies on pain of sanction that the players cooperate rather then defect; a *deus ex machina* appears on the sidelines of the game.

Robustness and Flexibility

A legal order will exhibit robustness inasmuch as its structure and the rights and rules embedded therein (including secondary rules of recognition, change, and adjudication) cannot be lightly modified or overridden in the face of changing demands. A right once established in law is not easily dismissed (though rights can and do undergo attrition with time). For, first and foremost, a legal system is an effective mechanism for protecting and promoting the rights of individuals and collective entities. *Morally,* a right is an interest with a special claim attached; *legally,* a right is anything to which a given role entitles an actor by virtue of occupation of that role (see above).[4] Few rights are totally indefeasible or inalienable. Other rights and the public interest will always have some claims.[5] However, if a right is to be violated, then a special justification is needed. An effective legal order will see to it that rights are not overridden lightly, either by governmental agents supposedly acting on behalf of the "public interest" or by economic actors pursuing their own profit.

Whether or not entrenched rights will contribute to robustness of *performance* in ecological terms depends upon the specific content of those rights. In practice, legal rights are often of questionable quality from an ecological perspective. So, for example, "use and enjoyment" rights have contributed to the tragedy of the commons. The rights embodied in Costa Rican law promote ecological destruction through deforestation. Polluters have often had the legally recognized right to discharge "reasonable" amounts of toxic emissions. Unrestricted rights to the enjoyment of common property are clearly devastating to the coordination demanded by ecological rationality. Unrestricted private property rights are little better; actors shielded by private property can and do engage in activities with considerable spillovers such as pollution. The rights of economic actors to engage in any activities of their choosing – rights secured in the capitalist epoch – inhibit collectively rational resource use. Private owners of a resource can pursue quick profits and exhaustion rather than sustainable yield.

Robustness of structure, then, will translate into robustness of performance only if the content of specific rights is appropriate. It is here that the flexible qualities of a legal order can come to the rescue of any

[4] Rights can come in the form of "side-constraints" upon actions to other ends (Nozick, 1974), as ends in themselves, or as licenses to act (Flathman, 1976: 6–7; Fried, 1978).

[5] Rights often conflict with one another; so the rights of the poor to the basic needs of life (Grey, 1976) may be inconsistent with the liberties of others; or the common law right against personal injury may conflict with the private property rights of a polluter.

incipient shortcomings on robustness. The content of legal rights can change with time, albeit very slowly.[6] Over recent decades, some small steps are discernible in the extension of the legal rights of individuals to a clean and healthy environment, and in the restriction of rights of access to common property resources (at least in the Western industrial world). One can imagine some further steps being taken toward the development of environmentally sensitive rights. Along these lines, one of the most interesting recent proposals concerns the establishment of legal rights for natural objects such as species, natural environments, and ecosystems. This proposal was advanced by Stone (1972) in the context of the 1972 *Sierra Club vs Morton* case in the USA, which concerned the intention of Walt Disney to construct a ski resort in California's Mineral King valley. Stone suggested that Mineral King valley should itself be granted legal standing. This proposition was accepted by Supreme Court Justice William O. Douglas.[7]

Despite a burgeoning "animal rights" movement (for example, Rollin, 1981), however, most legal scholars and moral philosophers are uncomfortable with the extension of rights to non-human entities. Whether or not natural objects have any *moral* rights remains a subject of considerable controversy among philosophers. For present purposes, though, that question is irrelevant. More important is whether the existence of such legal rights would have positive instrumental value for ecological concerns such as negative feedback and coordination. There is a precedent: clearly, corporations have no moral rights, yet legal systems in capitalist societies have found it expedient to treat them as "legal persons." Among other benefits, this treatment enhances the economic rationality of the market system. There is no reason why natural objects such as ecosystems should not be granted legal "personhood" in order to enhance the ecological rationality of human systems in their interactions with natural systems.

From a *very* instrumental and therefore expedient perspective, ecologically minded interests can take advantage of possibilities for changing the content of rights and rules. The limelight of occasional crises, whether major ones such as that of the "environment" around 1970 and the oil crises of the 1970s, or more localized ones such as Love Canal, Three Mile Island, Times Beach, Bhopal, and the breakup of the *Amoco Cadiz,* can be put to good use to facilitate the enshrining of

---

[6] Such changes will often be disguised as clarifications and interpretations of established rights.

[7] The Supreme Court as a whole ruled against granting standing to the Sierra Club – let alone Mineral King – in the dispute.

environmental values in law. Opportunities such as the Mineral King case can perform a similar function. Environmentalists can capitalize on any temporary public sympathy for ecological values that crises engender by pushing for legislation that will persist to structure legal and thereby social choices through ensuing periods of public indifference. Any more fragile form of social choice (such as polyarchy) might respond to such indifference by moving away from ecological concerns. In this respect, it is noteworthy that much of the 1970-era legislation in the USA continues to constrain would-be despoilers and depleters, even when those individuals reach high office in the executive branch of the federal government. Operating here is a kind of robustness of performance depending on some past expedient adaptations.

Negative Feedback

While law's main claims to ecological rationality lie in coordination and a combination of robustness with flexibility, a legal system can also promote the generation of negative feedback, in at least three ways.

First of all, a legal system allows actors to bring cases to its attention. Any (legal) individual suffering from the consequences of an environmental abuse can undertake legal action to seek redress from the perpetrator of that abuse. The possibility of any such action – and the probability of its success – will depend, of course, on the content of prevailing legal rights. So, for example, individuals suffering direct damage to their health and property and indirect damage to the protective (and, in the long run, productive) capacities of local ecosystems as a result of the burning of surplus straw in the fields of English farmers currently have no right of redress. But those same individuals can expect a hearing if watercourses in which they fish are polluted by those same farmers.

Second, a legal system can facilitate an injection of very high-quality negative feedback into debates over collective choices, irrespective of whether those choices are ultimately decided within the legal system or by some governmental mechanism. The forensic ideal is one in which protagonists make and respond to arguments in open fashion according to a set of neutral rules. The result is a dialectic of good argument in which all pertinent values are brought out. Thus one might expect better scrutiny of ecological values in a legal system than in ordinary polyarchical interaction. A court case offers the opportunity for ecologically minded interests (and their opponents) to develop cases, achieve publicity, and collect evidence in support of their claims.

Third, at least in the polyarchical context, negative feedback can be enhanced, and positive feedback inhibited, through the power of courts to curb special interests. It is far more difficult for such interests to develop cosy relationships with courts than with agencies or legislators. In practice, of course, access to the courts is itself highly skewed; one resource at the disposal of powerful organized interests is an ability to hire good lawyers and shoulder the expense of fighting cases. Nevertheless, courts can force the Japanese government to pay compensation to the victims of Minamata disease; can delay offshore oil lease sales in the USA over the opposition of oil companies and federal agencies; and can force environmental protection agencies to follow their legislative mandate more closely. A recognition of the courts' potential for curbing special interests offers additional backing to Lowi's (1969) proposal to replace corporate pluralism with "juridical democracy" (see above).

Resilience

It may be stretching a point, but legal systems also have some pretensions to resilience. While even friends of the law recognize that law as a problem-solving device is a fairly blunt instrument, a case can be made – if only on purely empirical grounds – that law can take a leading role in the rectification of disequilibrium conditions in man–nature systems. Certainly, most of the (modest) gains made in public action for environmental protection in Western polyarchies before the 1960s came through the courts. So, for example, the common law right to clean water of the owners of fishing rights in British rivers is long established, and accounts for the cleanness of British rivers in comparison with those of other industrial societies (Dales, 1968: 68–70). Admirers of the potential progressiveness of law often point to desegregation and civil rights in the USA, where it was the federal court system which took the initiative, eventually dragging the other branches of the federal government and state governments along with it.

Arguably, though, the leadership taken by the federal courts in regard to the civil rights issue in the USA is a result of the fact that all other channels of political influence were denied to racial minorities and their advocates. This generalization may be applied to environmental issues too: environmental groups make extensive use of the courts when they cannot be heard elsewhere in a political system. The case of Japan is exemplary here; the national government has been thoroughly insensitive to environmental concerns, but is occasionally prodded into compensatory action by the courts, as in the celebrated 1973 Minamata case (see Kelley, Stunkel, and Wescott, 1976: 189–91).

While a (somewhat thin) case can be made that law as a social choice mechanism can embody resilience, the strength of law's pretensions to ecological rationality rests ultimately, one suspects, upon coordination capabilities and an intricate combination of robustness and flexibility. Together with those qualities, any negative feedback, if somewhat haphazard, might be sufficient for law to stake a strong claim to ecological rationality. The case against has yet to be heard, though.

LAW: THE CASE AGAINST

Law in practice tends to be intertwined with other forms of social choice; indeed, it often provides the formal structure for them, as in the definition of the bundles of rights which can be exchanged in markets, or of the formal extent and limitations upon the powers of actors in a polyarchy. The content of law can be determined directly by governmental choice, and indirectly by the demands of the market system (both through its "imprisonment" of government and through the requirements of exchange systems for clear definition of the laws of property and contract). Therefore law is often a reinforcement of other forms of social choice rather than an alternative to them.

Moreover, legal systems are not insensitive to the distribution of power in the forms of social choice with which they co-exist. So, for example, in systems combining polyarchy and private enterprise markets, it is the more powerful interests which will have the greatest access to and influence upon the courts. In the Costa Rica case, powerful cattle and timber interests do quite well out of the existing system of rights and rules which promotes deforestation (see above). So the law may reinforce the positive feedback of markets and polyarchy (see chapters 7 and 9), instead of introducing any negative feedback of its own. In this context it is noteworthy that, at least until the 1960s, the judicial branch of government in the USA generally favored commercial and industrial interests in pollution cases (see Wenner, 1976: 141). The common law tradition of "personal injury" placed the burden of proof on the plaintiff in cases against polluters. Further, the courts developed the doctrine of "reasonable use," which excused harm done by polluters on the grounds of the benefits (economic or otherwise) of polluting activities (Wenner, 1976: 9).

Setting aside these problems of real-world legal systems, there are a number of features of even the best legal system which call into question any pretensions to ecological rationality.

Consider first of all one of the law's stronger claims: coordination. A court settles any dispute through reference to a piece of legislation, a constitution, a precedent, or common law. A legal order of any maturity will be highly differentiated, and hence will possess little consistent and uniform notion of the common good. Coordination through a consistent set of rules is not easily achieved. A legal system's response to ambiguity emanating from complexity and novel circumstance is to create yet more rules and finer distinctions.[8] Given that all laws must lay claim to generality, these finer rules must themselves be applied universally. This proliferation increases the likelihood of conflict between different rules, which in turn demands more rules of precedence. The net result is a detailed structure of rules which matches the complexity of the world it is attempting to cope with.

In its drive to refine, differentiate, and make itself applicable to as wide a range of circumstances as possible – a process sometimes referred to as "legalism" (Diesing, 1962: 140) – law inevitably addresses itself to the details of each specific case. Any new case is dealt with in the light of the formal language of statutes, their interpretation, and their precedents. The wider significance of a case is all too easily obscured, as judges will tend to decide cases on as narrow grounds as possible. So, for example, the decision to allow reprocessing of nuclear materials at Windscale in Britain was the subject of a judicial inquiry, but only energy needs and safety concerns were addressed in the inquiry's findings; the broader issues of a plutonium economy and the kind of society it demands were ignored (see Cotgrove, 1982: 83–4). In the ecological realm, the whole is more than the sum of its parts; different decisions have cumulative and synergistic effects.

This tendency for decisions to be made on as narrow grounds as possible means that different but related cases can be settled according to quite different principles. The rich variety of statutes and precedents upon which judges can draw is a recipe for disjointedness, fragmentation, and inconsistency of decisions. Thus one court may allow oil and gas development in a wilderness area on the grounds of clear legislative intent, and another may block the necessary transportation system on the grounds of its violation of the rights of local residents or wilderness users. The ensuing incoherent patchwork of decisions is hardly the coordination demanded by the non-reducible nature of ecological problems. Law has no device for achieving coordination under conditions of complexity.

Let me turn now to the other desiderata constitutive of ecological

---

[8] The comparable tendency of bureaucratic organizations to promulgate more rules in response to implementation failure was noted in chapter 8.

rationality. The robustness of law, while admirable in itself, has an unfortunate concomitant in a degree of rigidity sufficient to hinder (if not decimate) both negative feedback and resilience.

Law is, undeniably, rigid. The potential rigidity of common law is illustrated by the persistence of a rather perverse set of rights and rules in the Costa Rica case discussed at the beginning of this chapter. More generally, once a court decision has been reached – especially if it is at a high level within a legal system – it becomes enshrined as precedent, and provides part of the context for the resolution of subsequent cases. If the latter cases are clearly novel, then judges have considerable latitude and discretion in the precedents they choose to apply (see above). But if those subsequent cases are formally similar to the prior one, decisions upon them will be severely constricted. So, for example, a court may approve oil development off Alaska through reference to prior decisions about New Jersey, ignoring the totally different nature of the ecological problems in the two areas. One of the hallmarks of ecological problems is their variability across time and space. A legal system seeking uniform application of rules is hardly the device to promote sensitivity of human actions to local ecological conditions. The impetus of law to "treat like cases alike," while laudable in the abstract, is a recipe for insensitivity to spatial and temporal variation in the content of ecological problems. For example, burning low-sulfur coal in the American West is a very different matter to burning high-sulfur coal in Ohio, but both cases may be subject to the same emission control laws. Uniformity in the application of rules means insensitivity to negative feedback signals.

An equally important aspect of rigidity arises from the fact that, once on the statute books, laws take on a life of their own, becoming difficult to repeal or amend. (Indeed, this difficulty constitutes the reason why law is robust.) Therefore, even if a legislature has the best will in the world, is ready to take expert submissions seriously, and resists the pressure of special interests, it still risks freezing into law what may be only a transient consensus among experts (Ackerman and Hassler, 1981). In the face of human ignorance of ecology and changing environmental conditions, any such consensus may be a fleeting thing. For example, a concern with sulfur dioxide in the immediate vicinity of smokestacks dominated deliberations over the American Clean Air Act Amendments during the 1970s; subsequently, though, it became increasingly clear that sulfates and long-distance pollution (such as acid rain) posed a greater environmental threat than local sulfur dioxide (Ackerman and Hassler, 1981: 60–2). Congress enshrined the forced scrubbing of coal-burning power plant emissions into law as the appropriate policy instrument, despite eventual indications that alternatives such as coal-washing or incentives to use low-sulfur coal were preferable. Needless to say, any such process hardly constitutes a recipe for effective problem-solving (resilience) in an ecological context, and is an inbuilt impediment to negative feedback.

Most rigid of all are constitutional provisions which stand in the way of effective action to cope with novel problems. For example, proposals to establish systems of charges for pollution control have a number of attractions in comparison with extant regulatory regimes. However, any attempt to introduce them in a manner sensitive to local environmental conditions is likely to run afoul of the US Constitution's prohibition of indirect taxes which are not uniformly applied (Anderson et al., 1977: 113).

Negative feedback in a legal order is further inhibited – and distorted – because the language of law is the language of rights (and their corresponding rules). Therefore the only feedback admissible under a legal order concerns values capable of expression in terms of legal rights. This restriction is unfortunate for two reasons. First of all, not all moral rights that can be argued for have the status of legal rights. Rights to basic human needs such as subsistence, shelter, and education often have doubtful legal status; the rights of ecosystems, natural objects, and future generations usually have none at all. The process through which moral rights achieve the status of legal rights is highly uncertain. So some questions will simply not be defined in even the best legal system (see Rein, 1976: 46 and 60–1). The flip side of this particular coin is that some legal rights may exist which have no moral counterparts. So, for example, lawyers for Reserve Mining argued that the corporation made no profits, and therefore should not be expected to pay for the clean-up of its operations (Wenner, 1976: 100–1). This claim is credible only in the context of the rights of corporations.[9] The claim was eventually denied in court, but not before contributing to the massive delays (and thereby further pollution) in the case.

Second, even if all moral rights did have the status of legal rights, the content of feedback would remain unduly restricted. Not all values are readily expressed in terms of moral rights. In a legal order, values capable of easy expression as rights have a head start over those incapable of such expression. Falling into the latter category are values such as the protective capacity of an ecosystem, species diversity in a locality, and life support (not to mention non-ecological values such as economic efficiency). While such values are not totally ignored by law, they are likely to be lumped into a nebulous "public interest" category.

---

[9] In fact, Reserve Mining was itself under the joint ownership of two highly profitable companies, ARMCO and Republic Steel.

The result is that law converts all questions – however multi-dimensional, intricate, and subtle – into ones of competing rights. In so doing, a highly formal and artificial structure is imposed on problems, and some of the values at issue may be lost sight of. So, for example, "use and enjoyment" rights to a common property resource are more easily addressed than the overall condition of that resource and any ecosystem of which it is a part.

Law's rigidity and its restrictions upon the range of values admissible and the manner in which those values may be expressed hardly constitute the conditions for effective amelioration of any disequilibrium conditions in man–nature systems. As a problem-solving mechanism, law is extremely cumbersome. For it is, in the first instance, a mechanism for resolving disputes between identifiable parties, rather than for making policies or solving social problems. Generally, the legal system comes into play only when an aggrieved party brings an action *against* some activity (such as pollution) or proposal (such as a plan to open land to strip-mining). Law can only determine whether such actions or proposals are consistent with existing sets of rights and rules. Therefore courts cannot design or evaluate practices; they can only pass judgment on designers and evaluators (such as administrative agencies), and then only from a procedural perspective, not a substantive one (Jaffe and Tribe, 1971: 656–9). The deficiencies of judicial tribunals such as the British Windscale inquiry (noted above) are very obvious in this context. As Alexander Hamilton noted, the judiciary possesses "neither force nor will, but only judgment" (quoted in Rosenblum, 1974: 120). Law with purpose ceases to be law. If we seek to devise ways to stop the man-induced encroachment of seawater upon coastal wetlands, to reduce acid rain, or to slow the buildup of atmospheric carbon dioxide, then law is of precious little assistance.

A further obstacle to resilience is that the range of potential solutions available under a legal order is severely restricted. One aspect of the pervasiveness of the language of rights is that lawyers tend to see any positive measures that might be undertaken in terms of changes in the structure of legal rights. Such changes may have positive consequences – as are conceivable, for example, in the case of Stone's (1972) proposal to grant legal rights to natural objects, or reforms in the laws governing land title in Costa Rica. However, alternative actions which might prove more effective are pre-empted. Moreover, changes in the structure of rights, once made, are not easily adjusted; there is little room for trial and error.

Resilience is inhibited still further because legal systems are ponderous in their deliberations – more so even than polyarchies. The number of rules which can be applied to any given case is frequently extensive, and the amount of evidence which can be introduced to back competing positions can be weighty. The forensic ideal is one of unhurried and considered deliberation, such that bad decisions – and the making, through precedent, of bad law – are avoided. "Wars of attrition" between different sides of a case can all too easily result (see Ophuls, 1977: 194–5). Note, for example, the prolonged legal struggles by victims of asbestosis and thalidomide in the UK, of agent orange in the USA, and of "Minamata disease" in Japan. While cases drag on, ecologically damaging behavior can persist; so the Reserve Mining company continued to dump 50,000 tons of taconite tailings per day into Lake Superior for years after action was initiated against it (see Cahn, 1978: 90–6). Courts are already overburdened with cases in a number of polyarchies; and, as American experience attests, increasing the number of judges and lawyers in society may add to that load rather than diminish it. Therefore handing a substantial additional burden of social choice to such court systems could easily paralyze them (see Jaffe and Tribe, 1971: 662).

CONCLUSION

Law offers the promise of correcting some of the ecological abuses of markets, polyarchies, and administered systems. In particular, law can achieve a degree of coordination which, together with a combination of robustness and flexibility, could stand it in good stead. However, any claims to coordination fail to stand up to close examination, and the very factors making for robustness also contribute to a rigidity which severely restricts the scope of negative feedback and resilience.

11

Moral Persuasion

CHINESE PERSUASION

In 1949 China got a new regime and a new kind of social choice. Its new regime was (nominally) communist, and its distinctive kind of social choice was persuasion. The period between the revolution in the late 1940s and the demise of Mao (and Maoism) in the late 1970s saw substantial political turmoil, but one constant was Mao himself – and his belief that it was better to change a mind than to destroy it. Maoist ideology seeks control not through administration, but by convincing the masses of a number of basic principles which, having been internalized, they would use as a guide to their individual and collective actions. The party was to function as a teaching organization. Sometimes persuasion came in massive doses such as the Great Leap Forward, the Hundred Flowers campaign, and the Cultural Revolution. The content and extent of persuasion was highly variable, but the basic form persisted.

None of the campaigns of the Maoist era featured ecological concerns. To the contrary: a "grain first" exhortation promoted deforestation, conversion of land from sustainable pastoral agriculture to precarious cropland, and soil erosion (see Smil, 1984: 41). Mao continued to extol the virtues of population growth into the 1960s (only in 1971 did drastic population control commence). The Great Leap Forward sought increased industrial production, with no attention to environmental consequences.

The post-1949 period in China has seen substantial environmental decay. Among the worst problems are deforestation, desertification, soil erosion (with concomitant flooding and siltation in China's river basins), and water pollution (see Smil, 1984, for a catalogue). Severe air pollution exists in the larger urban areas.

China began to wake up to the magnitude of its ecological problems in the late 1970s. Recent years have witnessed an outpouring of information about environmental degradation and abuse. The official response has come largely in the form of exhortations to behave in environmentally sound ways – for example, the *People's Daily* sometimes publishes appeals to conserve resources. Public reaction to such appeals has been limited, for persuasion as a form of social choice in China has lost the power it had in Mao's day. Today, persuasion is just one instrument among many – just as it is in most other countries.

Nevertheless, one can imagine the application of persuasion with ecological content on a large-scale basis – whether in China or elsewhere. As we shall see, there are a number of Western advocates of such an approach to the resolution of ecological problems, either alone or in conjunction with other mechanisms.

SOCIAL CHOICE BY PERSUASION

Mahatma Gandhi warned us against "dreaming of systems so perfect that no one will need to be good" (Schumacher, 1973: 24). The story which has unfolded in the preceding four chapters suggests something very different: if men are not good, then markets, administered systems, polyarchies, and legal systems will fail, at least in terms of the ecological rationality standard. At this point, two options present themselves. One could either accept Hume's (1777) dictum to "design institutions for knaves,"[1] or, alternatively, one could explore ways of making people less knavish. Operations to affect the human motivations which are constitutive of social choice are the subject of this chapter.

A system directed toward the achievement of desired social ends through operations on the content of human motivations is termed "preceptoral" by Lindblom (1977). For Lindblom, a preceptoral system consists of a leadership elite instructing the masses in the interests of "moving toward centrally designed aspirations" (Lindblom, 1977: 56). Lindblom's analysis is grounded in observations of such mechanisms in operation in China and Cuba. For present purposes, this definition is unduly restrictive. I prefer to use the category of "moral persuasion." Moral persuasion allows for the possibility of aspirations inhering somewhere other than in the designs of an elite – for example, in the latent moral principles of individuals – and, concomitantly, appeals to values in addition to instruction in them.

In a political context, moral persuasion can take a number of forms:

---

[1] Hume commends this strategy on the grounds that it is fail-safe; people may in fact be virtuous, but it makes no sense to count on it.

education, propaganda, discussion, reasoning, linguistic manipulation, and exhortation. The agents of persuasion therefore include educational institutions, the media, religious institutions, and political leaders, activists, and organizations. In this form of social control, then, there is no threat of force, no sanction of law, and no place for the material incentives used in administered and market systems. Moral persuasion as a distinctive form of social choice retains a high degree of individual autonomy and volition.

Social choice through moral persuasion involves inducing individuals to engage in practices other than those that their unconstrained volition would lead them to. Assuming that people need little persuasion to do what is in their own immediate material interest anyway, preceptoral social choice must induce people to carry out actions and follow practices which impose a burden (such as sacrifice of material benefits) upon them. It is, then, the *moral* principles of individuals which are the target; that is, those guides for action which individuals hold over and above ordinary self-interest. An effective system of moral suasion must therefore be based on a sound theory of the moral sentiments that people hold, why they hold them, and how those sentiments relate to more egotistical impulses. Goodin (1980) notes that moral sentiments can be of three forms: the "prudential morality" of enlightened self-interest, under which people behave "morally" only in the anticipation of payoffs from reciprocating behavior on the part of others; the "internalized norms" of moral principles, entered as just another element to be satisfied in a utility function; and, most fundamentally, "seriously held principles." The latter constitute a very distinct class which individuals set apart from mundane calculations, a class so important that even to compare it with egotistic considerations is to corrupt it. A prime example would be a sense of patriotic duty in wartime. These "seriously held principles" constitute the only category of moral beliefs immune to tradeoffs and corruption by ordinary self-interested behavior. Therefore it follows that any *effective* system of moral persuasion will either appeal to or affect the content of this class of principles.

To be effective, and to hold any credence as a universal guide to action, persuasion must be of uniform content; and it must emanate from a relatively small group. The moral persuasion ideal is, in fact, close to the Utopia described in Plato's *Republic*. The Republic is ruled by an elite group (or individual) whose authority is that of superior intellectual capabilities. This "philospher–king" is more than a mere administrator, though. The state trains its people in virtue; all classes in the Republic then work toward the common good, as each individual puts his energies to use on the tasks to which he is best suited. The citizens of the Republic

152   *Evaluation*

are not coerced, nor are they automata; they are enlightened, educated people who accept the legitimacy of the ruling elite. In similar spirit, Jean-Jacques Rousseau considered it important that the state educate its citizens to work for the collective good (see especially Rousseau, 1972).

Moral suasion as a form of social choice should not, then, be confused with the unthinking compliance and appeals to the irrational side of human nature sought by mass persuasion in fascist systems (Lindblom, 1977: 54). It is active participation in problem-solving that is sought, not passive compliance. Moral persuasion may be contrasted too with the behavior modification recommended by some psychologists as offering solutions to social problems, including environmental problems (for example, Cone and Hayes, 1982). "Behavioral technology" consists of positive and negative reinforcement, punishment and extinction. Such "technology" is eclectic in the kinds of incentives used – which may range from legal sanctions to the provision of information – but it is unmistakably control and compliance which is sought. The individuals of the behavioral psychologist are amoral stimulus–response machines.[2]

Moral persuasion in Lindblom's "preceptoral" sense was the dominant form of social choice in Mao's China, and was applied to a lesser degree in Cuba from 1966 to 1970 (see Lindblom, 1977: 276–90). Elements of moral suasion can, though, be found in most other forms of social choice. In all cases, the relatively powerful may be discerned trying to influence the worldviews of the relatively powerless.

In market systems, persuasion takes the form of a constant barrage of advertising. Advertising is generally amoral in content, though it can shape the "market morals" (and hence the content of feedback signals) as described in chapter 7. Occasionally, one finds advertising by corporations on behalf of the market as an ideal – or, when the going gets tough, to promote moral (sacrificing) behavior such as energy conservation. In administered systems, the burden of control is lightened if subordinates can be convinced of the mission of the hierarchy and their obligations to it. Thus Kaufman (1960) describes how the US Forest Service successfully socialized its Rangers into a set of principles which can be applied to varied cases. Some Soviet bloc countries have seen official (if limited) exhortation that individuals should behave in environmentally sound ways (see Fullenbach, 1981: 93–5).

In polyarchies, suasion takes the form of political socialization by agents such as the school, family, religion, and the media. Indeed, a stable

[2] This kind of social control is not worthy of discussion as a social choice mechanism in its own right. Behavior modification can be expected to share the defects of administered systems discussed in chapter 8.

polyarchy needs the socialization of its citizens to produce the judicious mixture of deference to authority and confidence in participatory capabilities termed the "civic culture" by Almond and Verba (1963). In addition to their continuous and pervasive socialization processes, polyarchies occasionally experience large doses of moral persuasion in the interests of some very specific ends. Prohibition in the USA constitutes an instance of political activism (under the auspices of the Anti-Saloon League), together with legislation and (eventually) a constitutional amendment being brought to bear in an attempt to shift social values in a less sinful direction. The successes of civil rights policies in the USA are a result not just of the positive content of legislation and judicial decision, but also of the values of toleration propagated by those acts. To a degree, morality *can* be legislated. Individual efforts in energy conservation in the 1970s were more than a reaction to higher energy prices; there was an element of "social learning" in response to media publicity and the exhortations of political leaders (Lindblom and Cohen, 1979: 18), such that conservation became a moral imperative.

Suasion is less easily located in legal systems; one is tempted to regard the lectures of judges to hapless defendants in this light, but more significant is the "respect for law" which legal systems demand if habitual compliance is to be assured.

The fact that moral suasion may be discerned in other forms of social choice does not imply that moral incentives are simply devices which can be used as an adjunct to those forms as is expedient to produce compliant behavior. As Goodin (1980) recognizes, if governments use both material and moral incentives to the same end, then the purity and seriousness of the moral incentives are destroyed. The consequence is that individuals will simply remove those moral considerations from their decision calculus. Titmuss (1971) outlines this process in the context of policy for the medical supply of human blood. The British system of voluntary, unpaid donations relies for its effective functioning on the seriously held belief of large numbers of people that giving blood is morally right. It is impossible to sell one's blood in Britain. If one *could* sell one's blood – that is, if the system used material incentives too – then the act of giving would be cheapened, and the system of voluntary donation would collapse. Countries such as the USA where a market in blood is well established, cannot readily introduce a voluntary system, as the act of giving blood is seen as just another commercial transaction. Hence moral persuasion in any given area of activity really does have an all-or-nothing aspect which fits very uncomfortably with other forms of social control (though not necessarily with other forms of social *choice*).

154 *Evaluation*

THE APPEAL OF MORAL PERSUASION

Any decentralized system of social choice has considerable attractions in terms of both negative feedback controls (generally internal and diffuse) and problem-solving capacities which might translate into resilience. The moral persuasion ideal is one of a decentralized system of problem-solvers, free to engage in experimentation, criticism, and trial and error, but all committed to the same goals. If this sounds far-fetched, consider the way polyarchies organized their social choice in World War II: true, there were heavy elements of central planning, but the war machines could not have worked on administration alone. The committed problem-solving spirit is also akin to that perceived by some observers of Chinese agricultural and industrial enterprises during the period of Mao's ascendancy. Under Maoism, the energy and creativity of the masses is to be unleashed in a cooperative attack on problems.

In a market system, negative feedback fails ultimately because the logic of the market is sufficiently powerful to distort the content of signals in a manner antithetical to ecological concerns. Polyarchy is an improvement over the market inasmuch as it promises resilience through a problem-solving capacity which could be directed toward the attainment of equilibrium in the interactions of human and natural systems, even if that promise can expect frustration through the pattern of interest representation. What polyarchy and markets lack very fundamentally, though, is any effective coordination device. "Invisible hands" fail to turn the trick in the face of complexity and non-reducibility in the ecological circumstances of social choice.

Moral persuasion's appeal is that it can retain the negative feedback and problem-solving of decentralized social choice at its best (a "best" unrealized in markets and polyarchy), while simultaneously attaining a degree of cordination forever beyond the reach of market systems and polyarchies. Moral persuasion is a very visible hand, a hand which could convince people of a common good over and above their more egotistical impulses, to which their energies should be directed in concert. The prisoner's dilemma is resolved by a generally accepted belief that it is right to cooperate rather than defect. Even if any such "seriously held principle" is not universal, the payoff structure in the game can be affected through "internalized norms" sensitive to the interests of others. In this context, Axelrod (1984: 134–6) notes that cooperative solutions to the prisoner's dilemma are facilitated to the extent that individuals are taught to care about one another.

Coordination under moral persuasion is achieved in a centralized manner. However, the pathologies and rigidities of hierarchy, which destroy negative feedback, resilience, flexibility, and, indeed, coordination in administered systems, are avoided. While it is true that moral persuasion in social choice requires a cohesive leadership group to act as guardians of the common good, this group – whether Plato's philosopher–kings, Druids, the Chinese Communist Party, or Winston Churchill's Cabinet – has a simpler task than the leadership of a purely administered system. Instead of pondering courses of action, reaching decisions, and passing them to subordinates for implementation, a preceptoral elite need only comtemplate and propagate values and broad principles for action. The cognitive load on such an elite is therefore much lighter than in an administered system. That elite need not enmesh (and paralyze) itself in the intricacies of decision. Moreover, no hierarchically organized bureaucracy is necessary. One may need a chain of teachers intermediate between leaders and masses, but this chain can itself be organized in terms of moral persuasion (Lindblom, 1977: 57). In this context, the hostility to bureaucracy of Maoist ideology is noteworthy.

Moral persuasion as a form of social choice holds out the promise, then, of simultaneous achievement of coordination, negative feedback, and resilience. A central elite would simply see to it that individuals everywhere were versed in the principles of ecology and were aware of and committed to environmental values. From then on, the elite need do no more than keep a watchful eye open, as the detailed content of social choices would be made through the volition and action of ordinary people. So Western Alaska villagers could safeguard the ecosystem of the Bering Sea and its productive potential; New England foresters and Midwestern utilities could engage in a cooperative resolution of the acid rain problem; fishermen could be trusted not to abuse the common-property nature of fisheries; factories would choose not to pollute local rivers; everyone would turn thermostats down in winter and up in summer; and so forth.

Under this kind of social order, one could rely on "volition" to determine the detailed structure and therefore the content of collective choices. But moral persuasion could perhaps also be grafted with good effect onto other decentralized social choice mechanisms. In a market system, moral suasion could act as an antidote to defective negative feedback by leading people to demand environmental quality, goods produced by ecologically sound production processes, and rewards for their labor only from such processes. In a polyarchy, individuals and organizations would be constrained in their pursuit of private interest (and therefore in their generation of inappropriate feedback signals) by a conception of the common good; indeed, the status of "special" interests would be thoroughly undermined.

The appeal of moral persuasion is a very strong one, and a number of those minded to resolve ecological problems have fallen for its considerable charms, seeing changed social values as the key. So, for example, Henderson (1981) wants to accelerate a "conceptual" and "cultural" shift from the current "entropy state" to her "solar age"; Harman (1976: 141–6) commends morality in public life, responsibility in the private sector, and spiritual education generally; Tribe (1974) advocates a "spiral of process" involving both practical actions (such as legislation) and a continual rethinking of ethical premises concerning the environment; Ophuls (1977: 243) speaks of a "metanoia . . . tantamount to religious conversion," such that "we must direct our concrete political activities primarily toward producing a change of consciousness that can lead to a new political paradigm" (Ophuls, 1977: 223); Milbrath (1984) believes it incumbent upon an environmentalist "vanguard" to chart the path to a "new society" – and to convince others that the path is worth following. Even quantitatively minded systems analysts can find it in themselves to call for a change in public moral principles (Randers and Meadows, 1973: 271).

While most of these authors see structural change proceeding hand-in-hand with the value change they commend, there is also a school of thought which identifies a shift in general ethical principles as *the* key to the resolution of ecological problems. So, for example, White (1967) is unhappy with the ecological consequences of the pervasive Judeo-Christian tradition of human dominion over nature, and calls for a rethinking of Western values. Cahn (1978) explores the prospects for a wholesale shift in American social values in order to embed an environmental ethic in existing economic and political institutions – institutions he has no desire to change. Such a shift would be effected through legislation, considered responses to crises, education, and personal example (Cahn, 1978: 266–7). Miles (1976: 236) suggests that a "redeeming vision" involving a "profound change in values" is necessary to bring into being a civilization which can live in harmony with the limits to growth. Schumacher (1973: 79–101) believes that economic developments issue ultimately from the human mind, such that if minds are influenced by education in ethics and metaphysics then desirable social practices will henceforth ensue. More concretely, Sewell and O'Riordan (1976) argue that curricula to activate environmental consciousness should be established within educational systems. The experience of agricultural extension services in the USA indicates the potential for education beyond formal schooling (see Beer, 1982).

THE IMPLAUSIBILITY OF MORAL PERSUASION

The preceding discussion made no mention of robustness or flexibility, one of which (or an adequate combination of which) must be secured if a social choice mechanism is to lay claim to ecological rationality.

Consider first of all the possibility of robustness. Any system resting on moral suasion as a form of social control and social choice is in fact extremely brittle. The continued effectiveness of such a system depends not only on the continuing awareness, motivation, and communicative competence of an elite, but also upon the constant willingness of the masses to go along with the prescribed values. Subversion of those values is possible at both the top and the bottom of the system. Concerning the top, one need not be a cynic to note that individuals who achieve leadership positions in society are not always motivated solely by any conception of the common good, let alone by an ecological conception of that good.[3] Any elite coming to power through democratic means is likely to be compromised on the way up; any revolutionary elite, even if it does not have blood on its hands, is vulnerable to deflection from its original purposes. Juvenal's question resurfaces: "Quis custodiet ipsos custodes?"

One possible solution to this kind of elite vulnerability is bureaucratization. A purpose may be safer embedded in a bureaucratic organization than remaining at the whim of an elite group. Max Weber noted that the usual fate of "charismatic" authority is to succumb to bureaucratization when the charismatic leader leaves the scene. However, any purported robustness achieved in this manner means that moral persuasion has yielded to administration as the dominant form of social choice, with all the dire consequences for negative feedback, coordination, and resilience that administration entails (see chapter 8).

At the other end of the chain of persuasion, it is possible that the masses may prove less than compliant. So Cubans may be corrupted by materialism transmitted across the airwaves from Florida, and "bad attitudes" may resurface among the Chinese comrades. Adherence to the content of moral suasion will rarely be universal (even World War II polyarchies had their profiteers and black marketeers): how does one treat those who choose to behave differently, let alone organize themselves into opposition? In practice, preceptoral social control as carried out in

---

[3] The founding fathers of the US Constitution are sometimes described in a benign light. Even in this case, though, it is possible to interpret their motivations in terms of protection of personal property and interests (see, for example, Parenti, 1983: 60–75). It is also noteworthy that they followed Hume's advice to design institutions for knaves; perhaps they did not trust even themselves.

China and Cuba has involved an occasional resort to repression. Repression may be concealed under a veneer of "re-education" – for good reason, as repression is really the antithesis of the preceptoral or moral persuasion ideal. Yet it seems that, if moral suasion is to be fully effective and robust, then competing agents of socialization – such as churches, schools, voluntary associations, perhaps even the family – must be neutralized, if not destroyed (Lindblom, 1977: 57). Repression will, in fact, destroy the spontaneity and autonomy of actions which promise effective negative feedback and resilience under moral persuasion.

The prospects for robustness in moral persuasion are, then, extremely slim. Flexibility fares little better. Flexibility could perhaps be achieved by a steadfastness of *basic* purpose from top to bottom in the chain of persuasion, together with a spirit of open inquiry concerning the values instrumental to that overarching goal. The problem here is that any spirit of inquiry over values is unlikely to leave any basics sacrosanct. If adherence to the central common purpose is thereby diluted, then coordination fails too, and so does social choice through moral persuasion.

The whole question of coordination is, in fact, highly problematical under moral persuasion. A charitable view of moral persuasion's capacities of coordination would simply recognize a degree of incompleteness. Under this interpretation, moral persuasion requires elements of other decentralized social choice mechanisms (such as markets or polyarchy) to complete the coordination function. One could rest easy with those elements because the content of social choices will be based on ecologically sound individual preferences.

Unfortunately, as pointed out in chapter 1, institutional forms can produce outcomes which are largely independent of the ethical content of the preferences of the individuals participating in them.[4] Moreover, if effective moral persuasion really does demand the elimination of competing agents of socialization, then allowing the existence of any other forms may prove perilous.

Moral persuasion *per se* lacks any spontaneous, decentralized coordinating device. If elements of other decentralized social choice mechanisms cannot be trusted to perform the coordination function, then the burden falls ineluctably on centralized devices. Such devices may be imposed from above, or developed "spontaneously" from below.

Conceivably, the central elite or relevant lower level in the chain of

---

[4] Note, for example, the way free-rider problems inhibit coordination in the supply of public goods, or how the tragedy of the commons can occur even if all the relevant actors are knowledgeable and concerned (see chapter 3).

persuasion could see to coordination through the propagation of additional moral incentives. Such incentives would have to be very specific guides to action, for complex systems can behave in counter-intuitive ways, such that actions with the best of intentions can have negative effects. Consider, for example, the well-intentioned installation of electrostatic precipitators on smokestacks: one of the consequences may be an increase in acid rain, for the ash collected by these precipitators is alkaline. If incentives must be detailed, then an additional cognitive load is placed on the elite. In the face of that load, the elite could decompose each problem and exhort the appropriate social units to deal with the parts of it. Any decomposition of a truly complex problem is, though, likely to be a mistaken one (see chapter 8); moreover, decomposition introduces an additional set of coordination problems. Alternatively, the elite could choose to work with a holistic conception of the problem – and hence oversimplify it, with a concomitant risk of propagating the wrong moral signals. So, for example, an exhortation to rural peoples in the tropical world along the lines of "self-sufficiency in fuels" is generally an ecologically sound idea, but not if it leads them to burn dung and biomass – or convert biomass into alcohol – thereby diverting essential nutrients away from replenishment of the soil (see, for example, Jenny, 1980). The Maoist "grain first" principle may have made sense in terms of short-term food supply – but its undesirable ecological ramifications were massive. Misplaced exhortations of this sort have the potential to interrupt and distort negative feedback signals from natural systems. The complexity inherent in ecological problems can, then, be as devastating to coordination through moral suasion as it is to coordination through administration.

Clearly, though, any central elite shouldering the burden of coordination in a system of moral persuasion needs to do more than transmit simple messages like "think ecologically," and thereafter let spontaneous cooperation take its course. The elite must be knowledgeable about both ecological principles and the human values and practices instrumental to the achievement of desired ends. The cognitive burden on this elite remains, then, a huge one. In comparison with the demands upon the controllers of an administered system, all that is lacking is the need to oversee the operations of a bureaucracy. But even that advantage is partially offset by the additional requirement that an elite of moral persuaders have in its possession a sound empirical theory of moral sentiments (see above).

The leadership in a social choice mechanism resting on moral suasion remains in the mistaken image of scientific authority which backs the idea of problem-solving capacity and resilience in elite control in administered systems (see chapter 8). Any central elite in a system of moral persuasion is, then, an obstacle to effective problem-solving, and therefore resilience. Popper's (1966) first and most fundamental enemy of the open society is Plato, whose republic would rest on moral suasion by the kind of elite described here.[5]

At this point, resilience might be rescued by an elite deciding to rely solely upon the power of argument and fully open discussion. Hence we come full circle: open discussion and debate imply potential subversion of robustness and flexibility, through articulation of and adherence to diverse values. Moreover, any such openness will destroy the coordination that elite persuasion purportedly secures, for openness will destroy the elite itself.

Any elite in a system purportedly relying upon moral persuasion does, in fact, face a considerable dilemma. If, in the interests of negative feedback and resilience, the elite relies solely upon the power of argument, open discussion and debate will result. In such a situation no uniformity of purpose is possible, so the system is readily subverted from below, and no coordination in the face of ecological problems will be achieved. On the other hand, if the elite seeks to achieve coordination and robustness by backing argument with a heavier hand – say, with coercion or detailed instruction – then the system is no longer preceptoral, but merely authoritarian. In the Chinese case, this dilemma was never resolved in Mao's lifetime. There were wild swings between periods of open debate and criticism (such as the "Hundred Flowers" campaign of the late 1950s), witch-hunts against those with bad attitudes (such as the Cultural Revolution), and a few periods of relative calm, ordinary repression and creeping bureaucratization.

CONCLUSION

Moral persuasion in social choice in an ecological context embodies a set of paradoxes sufficient to disqualify its pretensions to rationality. If robustness is desired, then negative feedback and resilience are eliminated; any attempt to secure flexibility will at best ensure a failure of

---

[5] Maoists might see the "mass line" once practiced in China as a way to overcome the problem of elite fallibility and ensure widespread participation in problem-solving. Under the mass line, party cadres "summarize" the perceptions of the masses for the leadership, which then authorizes actions and transmits instructions back to the masses through the cadres. While not totally devoid of substance, the mass line fits uneasily with party control; further, it requires a prior commitment on the part of the masses to the principles expounded by the party leadership. On the mass line generally, see Schurmann (1968).

coordination; coordination itself can be achieved, if at all, only at the expense of negative feedback and resilience; and the latter two qualities can be achieved only if robustness and coordination are discarded. The outcome is a kind of impossibility result in which the five desiderata constitutive of ecological rationality cannot be achieved in combination, even if each could be achieved individually (which is itself doubtful).

The limited experience of real-world systems relying heavily on moral suasion to make social choices is not too encouraging. In Cuba and China preceptoral social choice has gone hand-in-hand with repression. In polyarchies and administered systems, moral suasion has been success-ful on a society-wide scale only in times of war; although, as the experience of the Vietnam war in the USA suggests, even that success is not guaranteed. Perhaps President Carter was aware of the circumstances of successful persuasion when he called for the "moral equivalent of war" to attack the energy crisis; the fact that his appeal failed miserably indicates that there are limits to the potential for moral suasion in the context of ecological problems. To adapt Gandhi's advice, one should not dream of men so perfect that systems no longer need to be good.

12

The International Anarchy

ECOLOGY GOES INTERNATIONAL

Ecological problems do not respect international boundaries. Indeed, some of the more severe and intractable contemporary ecological problems are truly global in their ramifications. The "greenhouse effect" of increasing atmospheric carbon dioxide, depletion of ozone in the stratosphere, and exhaustion of the world's fossil fuels fall into this category. Further issues, though not inherently global in their effect, have been consigned for resolution to the international community as a whole. Think, for example, of whaling, the future of Antarctica, and ocean-floor minerals (seabed mining has been one of the primary concerns of the United Nations Conference on the Law of the Sea).

Somewhat more common are regional ecological problems whose resolution demands concerted international action. Examples here would include protection of ocean fisheries against overfishing, oceanic pollution, and climatic change (such as that associated with desertification in Africa, or the proposed southward diversion of Siberian rivers). More common still are international transboundary pollution problems. Many of these concern air pollution – acid rain especially is blown across borders. In addition, several of the world's great river basins encompass different countries, so that pollution of rivers and lakes can be an international problem too. Transboundary environmental problems are especially acute in Europe, where a number of fairly small yet highly populated and industrialized countries are juxtaposed in close proximity. Further, many seemingly domestic issues prove on closer inspection to have an international dimension. For example, the extent to which France decides to rely on nuclear power may have an impact on both the global demand for fossil fuels and the likelihood of nuclear proliferation.

For better or for worse, the kinds of problems to which I have alluded will be dealt with (if they are dealt with at all) in the international arena.

Social choice in the international system will have substantial implications for the ecological future of humankind. The international system is important and pervasive enough in its ecological consequences to merit serious attention. And such attention it has indeed received from a number of writers, if rarely in the manner followed in this study[1]

For all their differences, the forms of social choice discussed in the previous five chapters share one important feature: the presence of a sovereign, governmental authority at the system level. The existence of the state is quite obvious in administered and polyarchical social choice. The state is less central, but nonetheless necessary, in markets, law, and moral persuasion. A market needs a governing authority, if only to enforce rules of property and contract.[2] A fully developed legal system requires some notion of binding authority.[3] And a system of moral persuasion needs central persuaders.

The international system, in contrast, has no governing authority. It is an anarchy. An anarchy is a decentralized social system lacking formal institutions at the *system* level. However, to characterize the international system as anarchy and evaluate it as such is, for present purposes, to obscure more than is illuminated. While the system is indeed a technical anarchy, I will henceforth avoid characterizing the international system thus because it is, in actuality, composed of a number of different forms of social choice, each of which may profitably be addressed in its own right.[4] For this reason, the technical anarchy of the international system will, for purposes of analysis, be broken down immediately into its component forms of social choice.

Before progressing any further in this analysis, a note on the actors in international social choice is warranted. There exists a "realist" or

---

[1] Many of these works are more adept at prescribing blueprints or Utopias than at engaging in serious evaluation (as opposed to sweeping condemnation) of the current international system. Among the better-known advocacies are those of the World Order Models Project (WOMP). For a survey of WOMPer thinking, see Mendlovitz (1975); for a critique, see Farer (1977). A more closely reasoned approach may be found in Caldwell (1984).

[2] This statement should be qualified by the recognition that there are spheres of market activity – notably the international system – lacking any government. Informal regulation in this kind of system will be discussed in this chapter. Domestic examples of market choice without some governing authority are less easily found.

[3] Again, an exception may be made in the case of informal rules; again, it is international law which constitutes the main exception; and again, this kind of mechanism will be discussed in this chapter.

[4] Also, I wish to reserve the more general term "anarchy" for a distinctive form of social choice which is applicable for the most part to the local social organization of individuals, as opposed to the international social organization of states and other collective actors. Anarchy in this sense will resurface in chapter 16 below.

"state-centered" school of thought which interprets the international system in terms of a collection of amoral, instrumentally rational, and self-interested states pursuing advantage as expedient (see, for example, Morgenthau, 1967). The realist interpretation is consistent with both a Hobbesian state of nature (Hobbes, 1946) and the ordinary-language pejorative association of anarchy with disorder or chaos. While this conception of the international system facilitates both interpretive accounts of the behavior of states and deductions and explanations concerning their interactions (Schelling, 1960), realism is something of an oversimplification.

A competing school of thought describes the system in terms of "complex interdependence" (for example, Keohane and Nye, 1977). This latter perspective notes the existence of international actors other than states: multinational corporations, churches, international banks, transnational non-governmental organizations, international interest groups (such as Amnesty International), political associations (for example, the Socialist International), scientific and professional associations, supra-national organizations (such as the International Monetary Fund and Organization for Economic Cooperation and Development), and so forth. Moreover, actors are viewed as complex collective entities, not simple unitary maximizers (Young, 1972; Allison, 1971). Corresponding to this variety of actors is a richer set of interactions transcending national boundaries; international relations are not just between states. The closer approximation of complex interdependence to reality is not, though, achieved without cost; for as the recognized complexity of a system increases, then so does the number of potential interpretations of that system.

This is not the place to take sides in the debate between these two schools of thought. I merely wish to point out that the relative importance ascribed to the forms of social choice extant in the international system (for example, coercive diplomacy versus rule-governed cooperation) depends upon the perspective one takes. Whichever school one adheres to, though, the list of component forms of social choice is the same.

It should be noted, too, that the international system is dynamic. While it is true that the basic elements of social choice have remained essentially unchanged since the inception of the modern states system (Young, 1978a: 245), the relative weight of these components can and does change over time. In some cases, this change can occur in dramatic fashion, as when a world war breaks out. Here, I do not wish to tie an assessment to the mix of different forms of social choice in the contemporary international system. This consideration provides an additional reason for disaggregation into components.

Still less do I wish to evaluate any *particular* international system, such as the one we have now.[5] The peculiarities of that system – such as the dominance of two superpowers, North–South inequalities, the existence of global energy cartels, and so forth – would, if addressed, serve to cloud the issue of the ecological rationality or otherwise of the constituent social choice mechanisms. Others more qualified have addressed the existing international system in such terms (for example, Kothari, 1974; Dolman, 1981).

THE FORMS OF INTERNATIONAL SOCIAL CHOICE

The international system, just like any domestic system, contains a mixture of a number of different social choice mechanisms. The following types are present.

(1) *Markets* International trade is generally organized in markets, of varying degrees of perfection.[6]

(2) *Administered systems* Administration in the international system was arguably more common in the days of empires and colonies than it is today. Yet organizations such as OPEC administer prices (sometimes), the International Atomic Energy Agency administers nuclear safeguards, NATO administers international military exercises, the bureaucracy of the European Economic Community administers a large domain in the member countries, the United Nations Environment Program undertakes some administration of environmental projects, and so forth. While administration is not particularly well developed in the current international system, some analysts are sufficiently enamored of its charms to prescribe centralized world government as a solution to global problems such as the threat of ecological catastrophe or nuclear holocaust (see, for example, Falk, 1971; Mendlovitz, 1975; Ophuls, 1977: 218–19).

(3) *Law* Despite its occasional Hobbesian proclivities, the international system is characterized by a body of accepted rights and rules which may be grouped under the heading of international law. While the authority of judicial institutions such as the International Court of Justice (and, in some cases, the United Nations) has increased over

[5] For a systematic (if somewhat terse) evaluation of the existing international polity in terms of criteria of peace, distributive justice, environmental quality, and stability, see Young (1978b).

[6] International markets in practice may be circumscribed by negotiated agreements between governments, by oligopoly, by monopsony, by coercion, and by the unilateral actions of national governments (for example, in imposing tariffs).

recent decades, compliance with the precepts of international law remains a largely voluntary matter.

(4) *Bargaining* Bargaining is a form of social choice under which interdependent actors participate in negotiations involving the making of offers and counter-offers, such that an outcome is secured when all the parties to a negotiation agree.[7] Bargaining has already appeared in the context of polyarchical social choice (see chapter 9). In the international system, bargaining among states and other actors is the major non-coercive, non-violent social choice mechanism.

(5) *Armed conflict* Armed conflict – organized warfare – is a fairly direct way of making social choices. Collective choices are arrived at on the basis of the relative military strength of the parties to a dispute. Organized warfare is a recurrent feature of the international system, though by no means unique to it.

(6) *Coercive diplomacy* Coercive diplomacy may be conceived of in terms of threats of violence, together with belligerent moves falling short of actual conflict, intended to produce acquiescence in an adversary. This mechanism of social choice therefore falls at an intermediate point between bargaining and armed conflict (see Young, 1978a: 254).

The international system, then, contains a very mixed bag of social choice mechanisms. At first sight, it would seem that each of the six mechanisms demands a full treatment in its own right. However, there is considerable scope for simplifying the task of assessment. First of all, markets and administered systems have already been dealt with. The discussions of chapters 7 and 8 were developed in the context of the operation of markets and administration at the domestic level, but the level of abstraction in those chapters was sufficient to ensure that the generalizations and judgments developed therein hold too for these mechanisms as they exist at the international level. Hence markets and administered systems merit no further consideration in this chapter. Second, coercive diplomacy is a mechanism intermediate between bargaining and war. One would therefore expect its ecological rationality or otherwise to be an amalgam of that of the latter two types. Third, inasmuch as international law possesses effective mechanisms at the system level which determine the content of rules and see to their

---

[7] More formally, bargaining may be defined as "a means by which two or more purposive actors arrive at specific outcomes in situations in which: (1) the choices of the actors will determine the allocation of some value(s), (2) the outcome for each participant is a function of the behavior of the other(s), and (3) the outcome is achieved through negotiation between or among the players" (Young, 1975: 5).

enforcement, it will share a great deal in common with legal social choice as dealt with in chapter 10. Inasmuch as international law involves decentralized compliance mechanisms and largely voluntary adherence to rights and rules, it needs to be addressed in its own right.

In sum, then, the two major tasks demanded of this chapter are assessments of the ecological rationality of bargaining and of war. International law must also be addressed insofar as there are aspects of it not covered under legal systems in chapter 10.

ARMED CONFLICT

One is tempted to dismiss organized warfare with an unequivocal condemnation as an ecologically irrational form of social choice. After all, modern conventional warfare is ecologically destructive. In some cases, destruction is intentional, as in defoliation using agent orange by the US forces in Vietnam, or in the more general possibility of environmental warfare. (For example, Iraq has used flooding to impede Iranian advances in their 1980s altercation.) More often, destruction is just an unintended byproduct of military activity. It hardly need be said that the ecological consequences of nuclear or biological warfare would be ecologically ruinous. Extinction of the human race and other species is a possibility in a general nuclear war (see Schell, 1982: 3–96). Further, the mere maintenance of a large military establishment imposes a considerable burden on the earth's natural systems. The modern military consumes vast amounts of metal, petroleum, and other mineral resources, and pollutes extensively too.

One should not blind oneself, though, to the ecological virtues of war. A case could be made in an ecological context that war *is* a negative feedback device; as pressure upon natural systems mounts (perhaps with human population growth), then competition for access to those systems (especially for productive resources) causes groups of humans to go to war, decimate their numbers, and hence reduce ecological stress.

This view of war as a negative feedback device may be a bit simplistic (if not necessarily wholly mistaken). A more subtle case can be made that war can promote both coordination and flexibility in social choice. In both cases, it is modern total war which can promote these qualities.

Consider coordination. The outbreak of war is frequently accompanied by individuals setting aside their self-interest and engaging in moral (sacrificing) behavior on behalf of the common good. Note, for example, the experience of the polyarchies in World War II. Such behavior can be accounted for in at least two ways. First of all, it could be a direct result of moral persuasion, itself facilitated by an atmosphere of common danger. Second, that behavior may be simply a result of enlightened self-interest, as people realize that war introduces dramatically increased uncertainty into their lives. Any person's future could be their own; therefore it is instrumentally rational for those individuals to contribute to risk-sharing arrangements such as the welfare state, or to public goods more generally (Dryzek and Goodin, 1986). While there is little to suggest that any of the public goods produced under such circumstances has ever included environmental integrity, one should not rule out that possibility.

Flexibility in social choice also can be secured by war – unfortunately, though, only for those who are thoroughly conquered. It is the thesis of Olson (1982) that crushing defeat has the effect of breaking up coalitions of special interest groups which, given time, would have succeeded in securing their own position at the expense of the general welfare. The successful post-1945 economic performance of Germany and Japan may be explained in these terms. Olson is interested only in economic growth (and therefore economic rationality), but there is no reason why such flexibility in social choice need be confined to the economic sphere.

War is also robust; it can and does occur in all kinds of contexts. A claim to resilience is less easily made. If the ecologically virtuous generally won wars and the ecologically oblivious lost, then one could expect war to move human systems closer to equilibrium with natural systems. Unfortunately, most environmentalists are distressingly peaceful. However, any tendency of war to decimate populations might also be an aspect of resilience.

None of these positive claims stands up too well to close examination. The negative feedback quality of war is of dubious validity, if only because most wars do not involve the death of a very substantial proportion of a country's population. The country which suffered the greatest population loss in the most destructive war in history (the Soviet Union in World War II) lost only 6 percent of its people. To achieve anything more substantial, nuclear or biological warfare would be necessary – a case of overkill, to say the least. Moreover, wars are more often started by the resource-rich than the resource-poor (except, perhaps, in internal wars of rebellion). There is little evidence to suggest that, as nations grow more prosperous, they also grow more peaceful. And war as social choice also involves a considerable degree of ecologically ruinous positive feedback. Arms races are an obvious example, but war itself often tends to beget still more war, rather than settling matters with any degree of permanence.

War can also directly inhibit feedback concerning the condition of natural systems. The military at war can with an easy conscience use the principle of "supreme emergency" to engage in behavior which under ordinary circumstances would be readily condemned. Natural systems can expect short shrift in wartime: note, for example, the massive pollution of the Persian Gulf as a result of an Iraqi attack on Iranian oil installations during the Iran–Iraq war in the early 1980s. Even in peacetime, environmental abuses such as the generation of large quantities of radioactive wastes and the ozone-depleting stratospheric flights of supersonic aircraft meet nary a murmur of dissent when undertaken by the military, even as they are widely condemned when contemplated or carried out by civilian concerns (Goodin, 1982: 220–41).

Coordination claims, too, are highly tenuous. It would seem that only total war can turn the trick here; and total war today would entail massive ecological destruction. Moreover, wartime can also engender attitudes and behaviors which are fairly expedient and self-interested rather than moral and concerned with public goods. Migratory birds have a hard time over the Lebanon of the 1980s, where a population armed to the teeth uses birds for target practice. Whales in the North Atlantic suffered a similar fate in World War II.

The most efficient way of fighting a war is through an administered structure. Both the military itself and wartime production are normally organized on a hierarchical basis. Thus war stimulates administered social choice at the domestic level, with all the dire consequences for ecological rationality detailed in chapter 8.

The summary verdict on war is, then, negative.

BARGAINING

Advantages

There are several ways in which bargaining might promise an ecologically rational negotiated order.[8]

First and perhaps foremost, bargaining is a means for achieving coordination in decentralized social choice. Bargaining can, of course, perform the coordination function in polyarchy too. The difference between bargaining in the international system and bargaining under

---

[8] Analyses of bargaining in the international system are hampered in that the theory of bargaining has been developed on the assumption that actors are instrumentally rational and self-interested (Young, 1975). Actors in the international system are internally complex and rarely fully integrated and purposive. In the discussion that follows I will draw on some of the more formal constructs of bargaining theory, but a note of caution is in order.

polyarchy is that international bargains tend to consist of formal negotiated agreements. For example, in situations which can be represented by the prisoner's dilemma game introduced in chapter 4, actors jointly commit themselves to cooperative behavior in the future. Any such agreements based on face-to-face negotiation should effectively prevent the prisoner's dilemma from arising thereafter – at least, if the parties do indeed abide by them. Bargains struck in a polyarchy can be more informal in nature, often shading into tacit understandings. Moreover, bargaining under polyarchy is only an ancillary means of coordination; processes of continual mutual adjustment across actors and decisions are more central.[9] Bargaining is the prime peaceful means of coordination in the international polity.

If a bargaining system is to yield any determinate outcome, then the existence of a "contract zone" is necessary. A contract zone constitutes the set of outcomes which all parties will prefer to the absence of an agreement.[10] In the language of game theory, a contract zone is based on the set of positive sum outcomes to a game. Any lack of agreement here may imply the continuation (or inception) of an uncoordinated Hobbesian state of nature (in, for example, widespread "defection" among actors exploiting a common property resource), or the coming into play of alternative social choice mechanisms, especially war or coercive diplomacy. The unattractiveness of the alternatives is often sufficient to create a sizable contract zone (Young, 1978a: 250). It is in part for this reason that substantial social order and coordination can be located in technically anarchical social systems.

Effective bargaining – like any solution to the prisoner's dilemma – is facilitated if the number of parties in negotiations remains small. As the number of actors increases, then so do transactions costs and concomitant difficulty in reaching a universally acceptable solution, and hence coordination. Fortunately, the international system possesses a far smaller number of actors than most other social systems, even if one includes non-state actors.

Effective coordination demands not only that agreements be reached, but also that the parties involved subsequently abide by any accords. Compliance in a decentralized system such as the international order is achieved through mechanisms such as the convergence of expectations around an agreement, the institutionalization of vested interests in an

---

[9] Such processes are not unknown in the international arena. To the extent that mutual adjustment does occur in the international system, coordination strengths and weaknesses will resemble those of polyarchy, discussed in chapter 9.

[10] The fact that there may be a large number of possible outcomes within this zone means that bargaining can often be a protracted process.

agreement, social pressures, fear of losing the trust of other actors, and enforcement by individual actors (see Young, 1979). The recent examples of individual-level enforcement would be the response of the British government to the Argentine invasion of the Falklands, or the threat of the US government to restrict Japanese access to fisheries in American waters should Japan fail to comply with International Whaling Commission prohibitions on commercial whaling.

A surprising degree of order can exist, then, in the international system. Such order is manifested most fundamentally in a substantial respect for sovereignty, a general understanding that promises should be kept, and a recognition that the use of violence should be limited (Bull, 1977). These three aspects of order may be described as "primary" values inasmuch as they are necessary antecedents to a coordinated pursuit of a wide range of other values in the international system (see Bull, 1977: 5).

Examples of international bargains struck and adhered to are not hard to find. International conventions cover postal and telephone services, air passenger transport, the testing of nuclear weapons, radio transmissions, European railways, the fisheries of the North Pacific, and so forth. Some anarchist writers see such agreements as proof of the possibility of "confederal" social choice – that is, coordination through cooperation among collective actors, without any central authority (see, for example, Ward, 1977).

Coordination of itself is not, of course, any guarantee of ecological rationality in social choice. I will consider now the extent to which the other criteria are met under bargaining.

Bargaining systems must rely for negative feedback upon the same kind of devices as polyarchy. In particular, the actors admitted to the negotiating table must bring with them some concern with the condition of natural systems in their locality or functional area of interest. Under polyarchy, such concern finds expression and response in a diffuse manner. Under bargaining, such concern must pass through the bargaining group, which is, in effect, a central "controller." Ultimately, that concern must reside in the bargain itself. So, for example, the governments of Norway and Sweden might bring their worries about the amount of acid rain they involuntarily import to a bargaining forum; any response to that concern will come only in the content of a negotiated agreement.

Bargaining's claims to resilience lie in the possibility of piecemeal construction of institutions and regimes (such as, for example, international fisheries conventions) through negotiations (see Young, 1978b: 208). With time, an institutional order could be built up, perhaps incorporating agreed-upon negative feedback and coordination devices. The record of transnational negotiation in the ecological realm (for example, to restrict access to common property resources, or to provide environmental public goods) is at best somewhat mixed. Transnational pollution such as acid rain in North America and Europe has yet to produce much more than rhetoric. International agreement on the Law of the Sea has proven elusive. Little global discussion has yet been initiated concerning the development of conventions or regimes for climate modification, the preservation of species (except for large, aesthetically pleasant, or mammalian ones), the "greenhouse effect," ozone depletion, or radioactive pollution (with the exception of nuclear test ban treaties). There are several items on the positive side of the balance sheet, though. A number of agreements on international fisheries have been negotiated (if at a regional rather than a global level). The existing negotiated regime for the Antarctic has some attractive features (Clark, 1984). A number of treaties restricting oil pollution on the high seas have been negotiated under the auspices of the International Maritime Organization. Regional agreements such as that covering the North Pacific Fur Seal (which has succeeded in conserving seal populations) have been negotiated in several cases. Bargaining on localized transboundary pollution takes place in regional organizations such as the US–Canada International Joint Commission and various international river basin commissions (such as those covering the Rhine and Danube). Symbolic agreements such as the 1972 Stockholm Declaration on the Human Environment might even pave the way for more substantive measures.

Shortcomings

Bargaining's strongest claim to ecological rationality lies in its attainment of the coordination desideratum. If bargaining is found wanting on this score, one suspects that its claim must fall by the wayside. Hence a contemplation of the deficiencies of bargaining may best commence with coordination.

Effective coordination under bargaining demands compliance with any agreements reached. The whole question of compliance with agreed-upon decisions is crucial in any decentralized system (see Young, 1979). It is far from unknown for actors to espouse support for the spirit and letter of an agreement while systematically violating its provisions. Such behavior should not be dismissed as mere hypocrisy; frequently, it is a consequence of the set of incentives with which an actor is faced.

There exists a category of circumstances in decentralized social choice in which non-compliant behavior is automatically punished, in that such behavior directly hurts the evader. Stein (1983) refers to such cases in terms of regimes for "common aversions." So, for example, a national railway in Europe might shift to a gauge different from the international standard – but it would incur substantial direct costs as a result; a country might decide to adopt a language other than English for its air traffic control – but the first major consequence would be expense and inconvenience to itself. International agreements to meet "common aversions" are, then, effectively self-enforcing. The "game" in no sense resembles the prisoner's dilemma, for there can be no positive payoff from "defection," irrespective of the behavior of the other parties. In passing, it is worth noting that anarchists arguing for the virtues of confederation tend to point to exactly these kinds of cases (for example, Ward, 1977).

Far more problematical are those cases where there is a positive incentive for non-compliance with agreements. Stein (1983) refers to such cases (again somewhat inelegantly) as "common interest" problems. Such problems can be characterized in terms of the prisoner's dilemma. Any agreement reached is highly unstable owing to the continuing temptation to defect by taking a free ride on the cooperative behavior of other actors. Take, for example, a negotiated agreement to control access to an international common property resource. All actors may agree that restrictions upon the use of the resource are necessary and desirable, and so may easily agree upon a regime for that resource. However, any non-compliant behavior with respect to the negotiated regime will not be punished automatically; in the first instance, it will be rewarded in the form of a greater "take" from the resource. The prisoner's dilemma resurfaces: all actors face a like set of incentives, all are aware of that fact, and therefore all stand ready to defect from the agreement, if only out of fear that others will do so first. At the international level, this kind of process can be observed in the case of ocean fisheries, whaling, nuclear proliferation, and sometimes in arms control agreements.

This latter "common interest" class encompasses virtually all international and global ecological problems of any interest or significance. A state's policy-makers might agree to a collective limitation upon global use of fossil fuels – but still see the burning of such fuels as necessary for its own development. A government might believe in non-proliferation and the treaties which embody that principle – but make an exception in the case of its own supreme necessity to develop nuclear weapons.

One further impediment to compliance in the international system which should be mentioned is that the notion of moral obligation, which can facilitate compliance and cooperative behavior in other decentralized systems, is lacking (see Young, 1979: 40). Collective actors cannot feel moral obligation; only individuals can. So, for example, governments do not always consider themselves bound by the agreements entered into by their predecessors. Note, for example, that the Reagan administration upon its accession in 1981 immediately withdrew US approval of the draft Law of the Sea treaty, believing, presumably, that the USA possessed an advantage in technology for the exploitation of seabed minerals sufficient to ensure that its companies would benefit most from a regime of free access.

This discussion should not be taken as implying that compliance with negotiated agreements is totally unrealistic in the context of international ecological problems. As noted above, there do exist a number of countervailing factors promoting compliance and cooperation in de-centralized systems. While compliance and coordination remain prob-lematical, bargaining cannot be sunk on this score alone. Negative feedback deficiency provides a suitable torpedo here.

Negative feedback under bargaining operates in a manner similar to that in polyarchy. Hence polyarchy's feedback flaws resurface here; as long as participating actors are self-interested in their motivation, feedback is negative with respect to political rationality, but positive with respect to ecological rationality.

Negative feedback under bargaining has some further flaws which (ideal) polyarchy evades. Recall that effective feedback under polyarchy requires an openness of access to political decisions, such that any category of people with an injury to rectify or an interest to promote can, by the mere act of organizing, expect to attain political influence and bend collective outcomes in their direction. Such ease of access is less conceivable under bargaining; "concern" is insufficient to secure a seat at the bargaining table, especially in the international arena. Greenpeace may be concerned about Russian and Japanese commercial whaling, but that organization cannot conceivably join the International Whaling Commission (IWC): any such concerns must be channeled through recognized participants – in the IWC case, national governments. This requirement adds further potential for distorting, if not obstructing, feedback signals.

At this juncture, one might attempt to rescue negative feedback by positing an openness of access to the bargaining table. Unfortunately, the greater the number of actors involved, the harder it is to strike a bargain, and so the less likely it is that coordination will be achieved. Therefore any attempt to promote negative feedback through expanding access will inhibit coordination, and vice versa. The protracted law of the sea negotiations – involving all the world's sovereign states – are indicative of the possibilities here. Moreover, any such delay is itself an impediment to the translation of feedback signals into collective practice.

Even when they possess only a few participants, bargaining systems are not noted for their lightning reactions to changing circumstances. That lack of responsiveness stems in large measure from the indeterminacy which the existence of a contract zone entails. But unresponsiveness is, too, a consequence of the fact that feedback must pass through a central "controller" – the bargaining group itself (see above). A highly clogged channel is likely. This centralization of feedback is a further reason why bargaining's feedback capabilities are inferior to those of polyarchy (under which feedback can be internal and diffuse).

Bargaining systems fail too inasmuch as they are neither robust nor flexible. Consider, first, robustness.

Bargaining systems are not robust because their operation depends upon a fairly restrictive set of conditions. To produce outcomes effectively and consistently a bargaining system requires a contract zone, some degree of common commitment to the three primary values referred to above, and a relatively small number of participants. Moreover, a rough equality of bargaining resources among the actors involved is necessary if social choices are to be negotiated rather than imposed by the more powerful actor(s). All four of these requirements can be problematical.

Consider first of all the necessity for a contract zone. In many cases, this requirement should be easily met. So, for example, *all* actors stand to lose as a result of unrestricted common property arrangements in international resources. A wide variety of possible agreements is likely to be preferred by all to the "law of capture" which would prevail in the absence of any agreement – assuming, of course, that those actors concur in the desirability of protecting the resource, be it the oceans, the atmosphere, the electromagnetic spectrum, or biotic diversity.

A contract zone will not always exist, though, as long as there remain some very fundamental conflicts of interest and values among the actors in the international system. The extent of those conflicts is, of course, contingent upon time and place, and so need not concern this analysis.[11]

---

[11] Current examples of such conflicts include those between the First and Third Worlds. From a Third World perspective, a concern with ecological values can look suspiciously like industrial countries trying to sacrifice the development of the Third World in the interests of resolving problems which are themselves the consequences of the depradations of the industrial world. There is, too, a considerable diversity in perspectives on ecological questions between East and West. The official Soviet position is that environmental problems are an inevitable consequence of capitalism, such that they can only be resolved effectively under socialism. So any participation in transnational bargains with capitalist countries will undermine the official Soviet position.

Suffice it to note that any social choice mechanism which can flourish only if those conflicts remain muted is a very fragile one.

Turning to the second requirement referred to above – shared commitment to primary values – respect for sovereignty, an understanding that promises should be kept, and a recognition that violence should be limited are widespread, but by no means universal. Governments can and do subvert and invade one another, break promises, and go to war.

The question of limitation on the number of participants has already been dealt with. Bargains between two parties are relatively easily achieved – though even here one should note the difficulties of the two superpowers in securing agreement on arms limitation. The difficulties multiply with the number of actors involved.

In an anarchic system, order and coordination can flourish only when there is a rough similarity of interests and capabilities. Clearly, the existing international system does not approximate that ideal. A considerable diversity of interest can stem from inequalities in the distribution of wealth and resources among countries (see Falk, 1975: 64–8). Once again, the picture is of a highly fragile state of affairs.

So much for robustness. Bargaining systems have some problems in flexibility too; for, once an agreement is negotiated, it is not readily modified or replaced (except through disintegration). Any process of amendment demands a further lengthy process. As circumstances change, then, international regimes and other negotiated arrangements – and bargaining systems themselves – can all too easily outlive their usefulness. Needless to say, institutions cast in concrete are hardly a prescription for successful adaptation to changing circumstances.

The rigidity inherent in bargaining mechanisms is reminiscent of that discussed in the context of legal systems in chapter 10, with the exception that negotiated agreements do not suffer from the detail and specificity of formal legal systems. On the contrary: vagueness and ambiguity readily enter into the content of agreements in response to any lack of commonality of interest. This situation can make for considerable difficulties when it comes to implementation (and hence coordination), as actors can interpret the content of agreements as they see fit.

The picture of bargaining which emerges is, then, one of a rigid and cumbersome form of social choice. At the international level, difficulties are exacerbated owing to the fact that participants in bargaining are rarely unitary, integrated actors. Therefore their need to achieve internal agreements adds substantially to the transactions costs of negotiation between them (Young, 1978a: 251). This is hardly a recipe for resilience. As the process of negotiation drags on, any disequilibrium which may have prompted negotiation in the first instance can go from bad to worse. Depletion of some of the world's species of whales to the point of extinction can be traced to this kind of problem.

In sum: bargaining systems exhibit severe difficulties in achieving coordination, have no effective mechanisms for generating negative feedback, are neither robust nor flexible, and are too cumbersome and rigid to lay claim to resilience.

INTERNATIONAL LAW

International law differs from fully articulated legal systems (discussed in chapter 10) in its possession only of primary rules (see Bull, 1977: 135). That is, there exist behavioral prescriptions for actors within the international system, but there is no specification of who or what may establish rules, modify them, or decide upon their priority in cases of conflict.

This absence of secondary rules and a central authority to enforce primary rules does not mean, though, that international law is a chimera. Most states (and other actors) comply with most international laws most of the time (Bull, 1977: 137). The predictions of Hobbesians notwithstanding, a considerable (if variable) degree of compliance on the part of actors in the system may be observed. So one finds widespread (if not universal) respect for sovereignty, adherence to the principle of diplomatic immunity, and, during wartime, compliance with the rules of war. Compliance in a decentralized system can result from self-interest calculations, fear of reprisals by another actor, felt obligation,[12] and socialization into a system of rules (Young, 1979: 31–4).[13] Over time, expectations converge around sets of rules. Even when international law is broken, the guilty party often feels the need to state overriding reasons – thus upholding the law even as it is broken (Bull, 1977: 138).

---

[12] Strictly speaking, states and other collective actors cannot feel obligations; only individuals within them can.

[13] The use of "third parties" as a form of social choice in the international system is described by Young (1978a: 256–8). This kind of mechanism involves the parties to a dispute seeking the services of a neutral third party, such as a court, mediator, or arbitrator. Arbitration is extremely rare in the international system. The use of courts (such as the International Court of Justice) is more common, and may readily be grouped into the category of international law as described here. Mediation does not currently exist in the domain of international ecological problems, though is has been applied extensively in international crises involving the possibility of organized violence (see Young, 1967). The prospects for environmental mediation will be explored at length in part III of this study. On the outlook for international environmental mediation, see Dryzek and Hunter (1987).

It is conceivable, then, that the functional equivalent of Hart's (1961) three secondary rules can operate in international law, if on a very decentralized basis. Inasmuch as international law does possess such devices (or even more formal institutions, such as the International Court of Justice), an assessment of its ecological rationality will be identical to the analysis of legal systems in chapter 10. Inasmuch as international law does *not* contain any secondary rules (or their functional equivalents), then the following circumstances will ensue.[14]

First of all, coordination will suffer simply because there will be considerable uncertainty concerning the existence and content of rules. One cannot readily comply with a rule if one is unaware of its content. Second, an absence of secondary rules of change means that any primary rules are likely to be extremely static. Such a static quality rules out negative feedback, flexibility, and resilience. Third, an absence of rules of adjudication means that endless dispute is possible as to whether a breach has in fact occurred, thus exacerbating compliance and hence coordination problems.

The net result of the special features of international law is, then, to subtract from the qualities of fully developed legal systems. Law of the latter sort received a strong negative judgment by the ecological rationality standard in chapter 10; international law can only be worse.

CONCLUSION

From the perspective of ecological rationality, the international system is clearly in bad shape. Its component forms of social choice – most notably war, bargaining, and international law – all receive an unequivocal negative judgment. Bargaining is worse than polyarchy, and international law can only be worse than legal systems with strong secondary rules. Recent history would tend to support this summary judgment; international problems such as transboundary pollution and general abuse of the ecosphere constitute perhaps the most intractable class of ecological problems.

---

[14] The following discussion draws heavily upon Hart (1961) and Bull (1977: 133–6).

13

From Evaluation to Design

Some major questions of judgment now loom large. The first question is a comparative one: Can the social choice mechanisms scrutinized in part II be ranked according to their performance on the ecological rationality standard? A second, related, question is, Do one or more of these mechanisms – alone or in combination – yield an adequate performance by that standard? The third question, contingent upon a negative answer to the second, is, How may one set about identifying ecologically rational innovations in social choice?

COMPARATIVE EVALUATION

The question of comparative performance merits a very guarded answer. The verdict on each of the seven mechanisms discussed in chapters 6–12 was, at best, "mixed." Clearly, none merits the unqualified approval which would stem from an adequate performance on four of the five criteria constituting ecological rationality (remembering that robustness and flexibility are substitutable). On the other hand, none of the seven completely lacks redeeming features.

The most dismal observation emerging from part II is that none of the seven mechanisms performs unequivocally well on *any* of the five criteria.[1] In each chapter a case against the claims of the mechanism at hand was carried on a point-by-point basis. This summary verdict implies that none of the more prominent social choice mechanisms in today's world can stake even a moderately good claim to ecological rationality. A cursory examination of the lack of progress toward the

---

[1] A possible exception arises in the case of robustness, especially that of markets, war, and law. However, robustness becomes moot without success on the two necessary conditions of negative feedback and coordination.

resolution of the more severe ecological problems in the contemporary world corroborates this assessment.

Despite the poor verdict each mechanism receives, though, the seven types do not stand condemned alike. The following comparative evaluations can be made.

(1) War receives the strongest condemnation of the seven. Armed conflict's tenuous coordination claims are far outweighed by substantial deficiencies on negative feedback and resilience.

(2) Bargaining is worse than polyarchy. These two forms of social choice are commensurable in that they share a number of important features, especially in their devices for negative feedback and coordination. Bargaining is, in effect, a clumsy version of polyarchy, lacking polyarchy's strength and variety of feedback channels. Moreover, polyarchy out-scores bargaining on resilience and flexibility. Only on coordination and robustness are the two roughly equal. Hence, to use the language of game theory, polyarchy "dominates" bargaining.

(3) Administered systems perform very poorly on all five criteria, with the possible exception of robustness. Even that robustness is of structure rather than performance, though. There is a sense in which administered systems are dominated by markets, polyarchy, and law. Markets clearly have superior coordination; polyarchy possesses better negative feedback, coordination, flexibility, and resilience; law has, one suspects, marginally greater coordination capabilities. The empirical evidence clearly backs the theoretical findings here; societies which emphasize administered social choice (notably those of the Soviet bloc) have made far less progress in coping with environmental problems than societies relying upon polyarchy and law. However, this apparent "dominance" warrants great caution. For markets, polyarchy, and law are not commensurable with administration in the sense that bargaining is commensurable with polyarchy. Administered systems make use of very different kinds of coordination and feedback devices than these three mechanisms. Hence, for example, if the performance of both markets and administered systems is adjudged "atrocious" on feedback capabilities, that result cannot really be interpreted as a "tie." For the quality of each on that criterion has meaning only in conjunction with the logic of the mechanism as a whole, especially as it relates to the other four criteria. Interaction effects in this context are sufficient to vitiate any attempt to keep a "score" by criteria. These reservations notwithstanding, administration is clearly a very poor mechanism.

(4) Markets, polyarchy, law, and moral persuasion all exhibit sub-

stantial weaknesses. Markets have a glaring lack of resilience, are extremely weak on negative feedback and coordination, and cannot readily undertake structural change. Polyarchy is not an unequivocal disaster by any criterion, but its promise on each is readily extinguished. Law's rigidity is sufficient to exclude negative feedback and resilience. Moral persuasion encounters substantial problems of coordination, robustness, and flexibility.

Any "winner" among the seven types of social choice would, then, be little more than the best of a poor bunch. Moreover, no such identification is easily made. For one cannot compute a straightforward summary measure from the "pluses" and "minuses" of each mechanism; the five criteria are not additive, and no common metric unites them. Any attempt at summation would fail to capture the richness of internal contradictions, "impossibility results," and interaction effects generally.[2] Therefore further energies devoted to the task of picking out a winner would be wasted. One must look elsewhere for ecological rationality in social choice.

Where, then, should one look? One possibility would be to examine *combinations* of the seven mechanisms evaluated in part II. Even if each of the seven may be deemed ecologically irrational standing alone, it may be that – in social choice as in ecology – a whole is more than the sum of its parts. Particular combinations might exhibit ecological rationality as an emergent property.[3]

While not inconceivable, such rational combinations are not particularly plausible. For if a mechanism *actively obstructs* any of the five desiderata that constitute ecological rationality – as opposed to merely failing to attain it – then that mechanism is not a good candidate for

[2] The difficulties inherent in producing a "winner" from any comparative assessment under conditions as complex as those present here may underlie the poor quality of contemporary debates on the relative merits of different forms of social choice – in the ecological realm and elsewhere. All too common are arguments on behalf of a mechanism based solely on the (real or imagined) defects of its alternatives. So socialist critics of the ecological shortcomings of capitalism present central planning as the "obvious" alternative. Defenders of markets base their case on some of the absurdities of Soviet-style administration. Liberal welfare economists derive the need for a purposive state from the shortcomings of markets. Conservative economists and public choice theorists turn this kind of reasoning on its head in calling for market-type strategies to correct for the defects of governmental mechanisms.

[3] Analogously, Hofstadter (1979) explores the way in which consciousness and intelligence in human brains can be synthesized from thoroughly unconscious and unintelligent components. In engineering, a frequent design problem is the construction of a reliable whole from unreliable parts (H. A. Simon, 1981: 23).

combination. Obstructiveness of this sort pervades the types of social choice discussed in part II. So markets "imprison" any governmental mechanisms with which they co-exist; administration in society constitutes an effective choking point for any negative feedback signals in circulation; polyarchy suffers a positive feedback cycle in interest representation and collective response; law, too, acts as an effective block to feedback; and moral persuasion is corrupted by the presence of any material incentives.

A more mundane obstacle is that exhaustive evaluation of combinations would not be particularly practical. The number of potential combinations of the seven basic forms – in various proportions, styles of interaction, and definition of spheres of operation – is astronomical.

One way of reducing that number would be to scrutinize only those combinations which actually occur (or have occurred) in reality. Thus one might look at the American mix of social choice, containing moderate doses of the free market, polyarchy, law, and administration; the Soviet system, with rather more administration and rather less of everything else; Western European systems, with the same elements as the USA but mixed in somewhat different proportions; various systems at the international level (for example, for ocean fisheries); and so forth. Again, though, the task would remain a massive one, and it would be arbitrary to confine oneself to the national level of analysis.[4]

COMMON FAILINGS

The social choice mechanisms covered in chapters 6–12 above might at first sight seem a disparate bunch. And certainly, the proponents of each mechanism make a great deal of its differences with the others. However, if we look at them in the light of their capacity to solve ecological problems, their commonalities are as striking as their differences. Commonality here is not just the simple fact of their near-universal failure as ecological problem-solving devices; more important is that all the mechanisms fail in one remarkably pervasive way. To return to the issue posed in chapter 2, they all tend to *displace* rather than *resolve* ecological problems.

First of all, the market is adept at displacing "scarcities" from one resource to another through relative price signals. Further, the market

---

[4] The comparative study of real-world polities in the light of ecological concerns remains, though, a worthy enterprise, deserving more attention than it currently receives. For some pioneer studies, see Enloe (1975) and Kelley, Stunkel, and Wescott (1976).

practice of discounting promotes displacement across time. Administration artificially decomposes complex problems, enhancing the likelihood that solution in one "subset" will have ramifications in other subsets. Polyarchy, too, decomposes complex problems – but on a largely random rather than planned basis. Law draws very tight boxes around each "case," and makes a virtue of resolving problems on as narrow grounds as possible. This tendency again ensures that only very selective aspects of a problem will be addressed, and broader ramifications ignored. Moral persuasion's ultimate reliance on the cognitive powers and prescience of some elite group renders it liable to the same fate as administration. Bargaining that produces determinate results must limit the number of parties and issues involved, again promoting artificial boundaries between problems and displacement across these boundaries.

The common failings of the world's major existing forms of social choice are substantial. Therefore any search for particular combinations of them (actual or potential) that would perform adequately on the ecological rationality standard involves a slender hope. Any mix is liable to yield only further opportunities for problem displacement. The next chapter will explore the possibility of a common underlying reason for problem displacement and associated defects in contemporary social choice.

DESIGN

Does there exist any actual or conceivable form of social choice of greater ecological rationality than the more entrenched mechanisms analyzed so far? If so, what would it look like? Would radical departure from the status quo be required, or would marginal modification of existing mechanisms suffice? To answer these questions, the focus of this study now shifts from the evaluation of established structures to the design of innovative mechanisms.

Design's central concern is the creation of form. In design, one canvasses, invents, crafts, develops and fine-tunes structures (or actions) with the intent of convergence upon desirable configurations (see Alexander, 1964; H. A. Simon, 1981; Dryzek, 1983d). Whereas evaluation is static, design is dynamic, treating the content of forms as variable. Design is central to all kinds of problem-solving demanding anything more than selection among fully articulated alternatives.

The first "innovation" to be addressed in part III requires only marginal reconstruction of some extant forms. This kind of social choice – the "open society" of free conjecture and criticism about proposals for collective action and public policy – is simply an idealized representation of the *best* kinds of problem-solving existing (or prefigured) in the major contemporary mechanisms. Thus in one sense the open society is a paragon of what already exists. Scrutiny of the capabilities of the open society is crucial for the larger flow of this study. In chapter 14 I will attempt to demonstrate that the open society, although an idealization, must fail as an ecological problem-solving standard (which is not to deny its substantial virtues in other areas). As the best that the contemporary mix of social choice mechanisms can offer, the open society confirms and highlights their common tendency to displace problems.

An alternative agenda for enhanced ecological rationality in social choice will be erected on the ashes of the open society. This agenda involves radical decentralization of the political economy and – more important – facilitation of "practical reason" in political life.

Part III

Innovations

14

Open Society

Unlike the social choice mechanisms discussed so far, the open society does not exist – and indeed, probably could not exist. On the other hand, all the extant mechanisms can be graded according to how well they approximate the widely praised open society standard. The open society does, indeed, promise substantially improved ecological problem-solving in comparison with these alternatives. Moreover, there are several senses in which the open society can be interpreted as the epitome of the best in problem-solving attained by these existing mechanisms.

My purpose here, though, is less to praise the open society than to bury it. Far from constituting part of an agenda for reform, the concept of an open society will be developed here as a foil for such an agenda. In other words, I will try to show how enhanced ecological rationality can be attained through rejection of some of the open society's central precepts. It is the *failure* of the open society to cope with ecological problems which is truly instructive. Let me begin by sketching the open society and how its precepts might be translated into social and political practice.

THE OPEN SOCIETY IDEAL

The most effective form of collective human problem-solving may currently be located in scientific communities, inasmuch as these communities are governed by free and open conjecture and criticism. It is on this conception of science at its best that Popper (1966) models his open society of problem-solvers. Popper himself and Popperians such as Campbell (1969) and James (1980) postulate an "experimenting society" of "piecemeal social engineering" in which limited social scientific knowledge informs and is tested by self-conscious collective choices (such as public policies), conceived of as experiments.

Both professionals (such as social scientists) and laypersons have roles in the open society. Proposals for collective actions (conjectures) and criticisms both before and after the fact can come from either quarter. Social scientists have a special role to play in analyzing – and, if possible, foreseeing – the unintended "secondary" effects of proposals or actions. In public policy, those secondary effects often dominate intended primary effects (Bauer, 1969). If secondary effects are to be captured with any fullness, then social science analysis must be admitted from *any* disciplinary perspective. But such analysis, however broad-ranging, is not of itself sufficient. Collective choices have multiple and diffuse effects upon different categories of people; only the individuals affected will have a relatively full knowledge of the content of those effects. Hence collective choices must always be open to criticism from ordinary people. These people may be expected to adhere to a wide variety of interests and values. In contrast to the situation that can prevail under polyarchy, though, arguments based on those interests and values must bear public scrutiny in the open society.

Public judgment over experts is desirable even in the context of highly technical policy problems. Think, for example, of the issues surrounding nuclear energy, toxic wastes, or climatological change. In such cases, the relevant professionals could debate one another before an informed and concerned (but non-expert) audience (see James, 1980: 172–3). Proposals for institutional innovations such as "science courts" are pertinent here (though Popperians might be unhappy with the legal formalism to which science courts could be susceptible).

Bias, prejudice, and sleight of hand in argument can be most effectively exposed in public discussion. The open society is concerned only with the generation of objective knowledge – that is, knowledge independent of its subject, or group of subjects. Knowledge in the open society inheres in the public realm, independent of the perspectives (and idiosyncracies) of individuals.

Here, it would seem, is a very seductive recipe for effective problem-solving in social choice; a form that does, indeed, rely upon feedback for its effectiveness, and which holds a clear promise of resilience. The more uncritical admirers of the open society might at this point proceed to devote themselves to the practical task of designing institutions for the promotion of "openness." Thus McClosky (1983: 158) suggests that "The only realistic, feasible avenue to ecological political reform is through the political institutions of an open society that respects human rights."

It is in this spirit that James (1980) commends institutional designs which strengthen the control of elected representatives over permanent officials, and simultaneously establish "overlapping loops of information and control" to make both elected and non-elected officials accountable to the public (James, 1980: 158–76). Depending on the issue area in question, the latter role could be played by organizations such as neighborhood associations.

The institutional proposals of Levine (1972) are also interesting here.[1] Central to Levine's program for reconstruction is the idea that competition between separate government agencies operating in the same policy area will improve performance, as bureaucratic rivalry will lead to mutual criticism and thereby to a sharpening of actions and proposals. Thus the Office of Economic Opportunity added impetus to the US federal government's "War on Poverty," following its creation in the mid-1960s, not just through its own efforts, but also through its effect on other agencies with anti-poverty responsibilities. Similarly, the establishment of specialized anti-pollution agencies in many Western countries around 1970 arguably led to more rapid and effective attacks on environmental problems than could have been accomplished through existing agencies.

Levine suggests too that policy-makers can sometimes promote problem-solving by creating (or enhancing the political power of) a constituency for a particular set of values, through political or legal means. That constituency will then engage in critical oversight of governmental activities. Think, for example, of the effects of the advances in legal standing achieved by racial minorities and environmentalists in recent decades in the USA.

In a more specifically ecological context, Jones (1975) presents some proposals with clear open society overtones (if less clear concrete implications). Jones's study of air pollution policy in Pittsburgh reveals a guarded satisfaction with a polyarchical process which could, however, achieve substantially greater instrumental rationality if the public were more informed of the technical aspects of the issues at stake. Thus invigorated, anti-pollution policy would take the form of bold but intelligent incrementalism, straining at the "outer limits of capabilities" (Jones, 1975: 307–8).

Jones's suggestions in their turn seem to mirror the intent of two major pieces of US environmental legislation. Both the National Environmental Policy Act of 1970 and the Technology Assessment Act of 1972 provide for the systematic generation and dissemination of information bearing upon policy decision, the unspoken assumption being that greater and more widespread knowledge enhances rationality in the policy process.

---

[1] Levine, unlike James, is not a conscious Popperian acolyte.

THE OPEN SOCIETY AS PARAGON

The most promising devices for the achievement of negative feedback and resilience in social choice identified in part II of this study reflect, however imperfectly, the processes of free conjecture and criticism attained by polyarchy and moral persuasion at their (unrealizable) best. Negative feedback and resilience in polyarchy are ultimately frustrated by three factors: the positive feedback of political rationality, any "garbage can" tendencies in decision (solutions chasing problems, rather than vice versa), and the cumbersome nature of interactive decisions. Political rationality (based on the mollification of special interests) and garbage can tendencies are both features of polyarchy which will, by definition, be absent in the open society ideal to which polyarchy is the nearest real-world social choice approximation. (Cumbersomeness is less easily dispensed with.) Turning to moral persuasion, negative feedback and resilience are inhibited by detailed instruction, bureaucratization, or repression arising from any central elite's attempts to secure coordination. Absent such tendencies, what would remain again resembles an open society.

The open society is a paragon, then, in the sense that it improves (by two of five criteria) on the best that extant social choice mechanisms have to offer. However, it is not only polyarchy and moral persuasion which prefigure the open society here. Let us consider the remainder of the mechanisms dealt with in part II in the light of their problem-solving capabilities.

Hayek (1979: 65–97) describes the *market* as a "catallaxy" of individuals pursuing their private ends under conditions of highly imperfect and fleeting knowledge dispersed across persons. To Hayek, the genius of the market is its ability to solve problems under such difficult conditions: "Competition is thus, like experimentation in science, first and foremost a discovery procedure" (Hayek, 1979: 68). The market can be interpreted in this light as a kind of open society in which information is simplified and transmitted via price signals, rather than verbal or written communication. Individuals – alone and in combination – can be efficacious problem-solvers in such an environment; if they are not, the market will punish them.

Administration is subject to a number of problem-solving pathologies attendant upon hierarchy, authority, and bureaucratic structure. To the extent that these tendencies are eliminated, compartmentalized problem-solving would resemble that of the open society. It is noteworthy that effective problem-solving sometimes occurs in an administrative setting when individuals succeed in ignoring or dissolving hierarchy and barriers between organizational units (see chapter 8; also Levine's proposals as discussed above). Further, administrative reforms such as matrix organization supplement hierarchy with relatively "open" (but still compartmentalized) problem-solving groups, organized on a transient basis around particular tasks.

Law is at first sight too constrained in its deliberative structure to allow "open" problem-solving. Yet the forensic ideal is one of free interplay of arguments – if based on pre-ordained and highly restrictive rules and value positions. If such restrictions were lifted, then open society problem-solving – compartmentalized on a piecemeal basis – would stand to benefit.

Bargaining too is highly constrained and cumbersome. Nevertheless, bargaining can go beyond the articulation of positions and splitting of differences to function as a problem-solving device. As such, it enables articulation and criticism of proposals to ends favored by one or more of the parties involved.

As problem-solving procedures, then, the social choice mechanisms dealt with in part II are efficacious to the extent that they resemble the open society, pathological to the degree that they do not. Yet there remains a deeper sense in which the open society epitomizes good problem-solving in contemporary social choice.

To state a thesis at its boldest, the open society is a paragon of the kind of reason prized by Western society. Such reason is instrumental and analytical in character. Phenomena are understood and problems structured through disaggregation into their component parts. Based on this disaggregation, actions are devised and effected in pursuit of essentially arbitrary ends.

Analytical reason is exemplified in the scientific community, which is, of course, Popper's model for the open society. Now, there is no deterministic causal link from the sensibilities of science to the structure of social choice – clearly, markets and administration have been around longer than post-Enlightenment science. Nevertheless, the kind of rationality embodied by science is widely recognized in contemporary Western societies, and social organization can therefore be *justified* on the basis of this conception of rationality. Moreover, the problem-solving capacities of scientific rationality are of obvious heuristic utility to reformers seeking to embed like capacities in social choice.

A subtle historical dynamic may be operating here. Institutions prove effective in solving problems to the extent of their "openness," and are retained on that basis. Ineffective, "closed" institutions should fall by the wayside – as long as society's members are (instrumentally) rational enough to distinguish the good from the bad, and act upon this distinction. This process should constitute a self-reinforcing spiral: the more openness in society, the more opportunity there is for individuals to exercise and sharpen their critical faculties, and therefore the greater is the likelihood that they will choose open institutions.

Thus a link can be established between the dominant kind of reason in society and the structure of social choice. It is more than coincidence that the open society represents the major contemporary social choice mechanisms at their problem-solving best. The open society is indeed a paragon, but, as we shall see, a paragon that fails in an ecological light.[2] The open society has its attractions, but the special factors in the ecological circumstances of social choice make some demands it cannot meet.[3]

ECOLOGY AGAINST THE OPEN SOCIETY

Ecological rationality in social choice requires practice that is "symbiotic" rather than instrumental in character (see chapter 4), for that intelligence must cope with temporal variability in the circumstances of choice, and must mesh with the spontaneous self-organizing and self-regulating qualities of natural systems. While the open society may constitute the ideal conditions for the exercise and refinement of instrumental rationality, it may falter when it comes to more symbiotic practice. Consider why.

For Popper and his followers, the model for the open society is the scientific community and the sensibilities it fosters. Western science has, though, restricted the scope of its inquiry to certain kinds of questions. That science is, as noted above, *analytic* in its methodology. Complex phenomena are apprehended through decomposition into their

---

[2] In recent years a number of speculative works have appeared which attack what is characterized as a Cartesian/Newtonian/reductionist/mechanistic society, and commend a transition to a more holistic/organic state. For one of the more lucid examples of this genre, see Capra (1982). While my own analysis too takes issue with currently dominant kinds of reason, it is far less ambitious than these accounts of Gestalt-switch.

[3] It is occasionally recognized in other contexts that different kinds of problems may demand different sorts of social choice. For example, E. F. Schumacher believes that small is generally beautiful in social organization, but that bigness also has its place: "For his different purposes man needs many different structures, both small ones and large ones, some exclusive and some comprehensive" (Schumacher, 1973: 65–6). Public choice theorists recognize that questions pertaining to allocative efficiency in the supply of public goods are best decided through mechanisms embodying unanimity as a decision rule, whereas distributional issues are better resolved through some form of majority rule (see Mueller, 1979: 207–26).

component parts. Thereafter, invariant statements about relationships between elements – scientific laws – are sought. Positivists believe that such laws can be verified; Popperians demur, arguing that propositions can only be falsified. However, successive attempts at falsification are seen as weeding out errors, hence "corroborating" or enhancing the truth-content of theories. Moreover, the very possibility of falsification is enhanced by decomposition into ever-smaller parts. Thus Popper can speak of "objective knowledge" in science (Popper, 1972), just as the positivists do.

To use a terminology associated with Aristotle, the realm of the scientific – for both positivists and falsificationists – is *theoria*, discovery of the invariant nature of the cosmos. Such understanding then yields the possibility of manipulation of causes to produce predictable and desired effects. While Popperians accept the possibility of error in those manipulations – because theory can only be in the process of corroboration – it is exactly such manipulations which would occur as the dominant form of social practice in the open society. Popperians equate reason with *instrumental* reason – that is, any other kind of practice is simply irrational, because its theoretical backing cannot be falsified.

Practice based on *theoria* is what we currently refer to as technology.[4] Practice in the form of technology and instrumental reason may be adequate in many settings (think, for example, of the vaunted successes of technology, such as the Apollo Project). But the special conditions prevalent in the ecological realm detailed in chapters 3 and 4 cast doubt on the adequacy of instrumental reason, and hence on the open society which raises that form of reason on a pedestal.

First of all, spontaneity and variability in ecosystems means that they often defy technical knowledge of cause and effect. An analogy may be drawn here with human social systems. Aristotle saw the realm of politics in terms of practical reason *(praxis),* meaning the cultivation of virtue in members of the *polis*. In politics, as in all human social systems, one is dealing with entities (human beings) capable of autonomous, goal-directed action, and therefore not readily subsumed and manipulated under scientific "laws" (a "science" of politics is of fairly recent vintage). Attempted manipulations of the latter sort underlie many of the failed governmental social programs of recent years in Western societies.[5] Human social and political systems have

---

[4] Technology based on scientific *theoria* is, in fact, a phenomenon which arose as recently as the nineteenth century. Prior to that era, the realm of *techne* was that of arts and crafts, an arena of knowledge but not scientific knowledge. See McCarthy (1978: 1–4).

[5] See Nelson (1977) for one catalogue of failures.

intentional or teleological components, and emergent properties at higher levels of organization. Such systems therefore defy reductionistic, analytical understanding, in themselves and – especially – in their interactions with ecosystems.

Ecosystems are not teleological (see chapter 3), but they are self-organizing, self-regulating, and characterized by emergent properties. Ecology is not a predictive science, only an explanatory one. Therefore the kind of practice appropriate to our dealings with them is more akin to Aristotelian *praxis* – involving purposeful actions and variable, contingent relationships – than it is to technology or instrumental reason. I do not mean to suggest that instrumental rationality is totally lacking in potential in human dealings with ecosystems; merely that it can be only part of the story. Technology and instrumental reason yield an image of control over nature: "resource management," or "environmental engineering." Ecological engineering has some very severe limits (see chapter 4), for ecological systems are often largely immune to the degree of control that instrumental scientific understanding requires. Think, for example, of the unintended and unforeseen effects of introducing synthetic chemicals or exotic organisms into ecosystems. Contemporary debates over the release of genetically engineered artificial organisms into the environment – for intendedly benign purposes such as controlling oilspills, or protecting crops against frost – suggest that even ecologists have little idea what the ramifications of any such releases will be. Given limitations of this sort, releasing artificial organisms into the environment makes little sense even on a piecemeal basis, for such action would assume, perhaps erroneously, that we would be able to engineer an escape from any undesirable secondary consequences.

The general difficulty here is that in an ecological context these secondary consequences will be both massive and unforeseeable (because ecology is not a predictive science). Moreover, attempted solutions to secondary problems may *themselves* yield a proliferation of still more "tertiary" problems. Therefore the net result may be an expanding range of unsolved problems, because our piecemeal problem-solving capabilities cannot catch up. Ehrenfeld (1978) outlines the dynamics of this kind of process in cases as diverse as the use of antibiotics, the application of chemicals for pest control, and large-scale ecological engineering projects such as the Aswan Dam.

A more intelligent approach to human interaction with ecosystems would recognize the potential for meshing with the autonomous and spontaneous actions of those systems. Ecologically rational practice involves intelligence *with* rather than control *over*. Symbiotic reason suggests an incomplete distinction between (human) subject and (natural) object.[6]

If the preceding discussion seems somewhat abstruse, the second flaw in instrumental reason and the open society whose ideal it is should be somewhat easier to comprehend. Recall that one of the central features of the open society is that proposals or conjectures can be criticized or evaluated from *any* perspective. If one equates "perspective" with "normative stance" – a fair equation in social choice and social practice – then clearly "anything goes" in terms of normative positions (except positions opposed to the very idea of critical reason and the open society). In this sense, the open society stands squarely in the tradition of moral relativism which has pervaded Western society since the Enlightenment.[7] Popper himself dismisses any attempt to impose a unity of purpose upon society as smacking of "historicism" – the contention that there exists a next, higher, stage of historical development toward which society should be moving.

What this restriction means is that any process of a community of individuals engaging in norm-formation (or adjudication, or discussion) is consigned to the realm of the irrational. Norms cannot be falsified; therefore they are beyond the reach of reason. Hence ecological values have exactly the same epistemological status in the open society as a commitment to monetary greed, or vegetarianism, or maximum sexual activity. A society governed by instrumental rationality can treat purpose only as arbitrary.[8]

All information in the open society resides in the public realm (see above); therefore, given a plurality of values and interests, it can be put to any purpose. That purpose may, indeed, dismay the progenitors of

[6] Some further very strong parallels with action in the context of social systems could be pursued here. So the "behavioral technology" which some psychologists seek to apply to the resolution of social problems (see Cone and Hayes, 1982 for an environmental policy example) is akin to ecological engineering, and equally misplaced. Habermas (1971: 310) sees any social scientific laws as capturing nothing more than "ideologically frozen relations of dependence." Habermas identifies the proper place for instrumental reason and technical control as the natural sciences. In contrast, the social sciences should, for Habermas, be governed by an "emancipatory interest" in the improvement of human existence through the raising of the consciousness of individuals. Such an interest bears a clear resemblance to Aristotelian notions of *praxis*. My suggestions indicate that ecology may bear more in common with the *praxis* of the social sciences than with the manipulative spirit of the natural sciences.

[7] See MacIntyre (1981) for a history and critique of this relativism.

[8] Friends of the open society might at this point see a wide variety of "biases" "cancelling out" in the aggregate (for example, Diesing, 1962: 179; James, 1980: 173–4). But the resultant is only *politically* rational: from an ecological perspective, biases and errors may just as easily accumulate and reinforce one another.

information. Ehrenfeld (1978: 248) describes the case of a marine mammal biologist deeply concerned about declining whale stocks. That biologist, acting under open society imperatives, chooses nevertheless to publish his research on the geographical distribution of certain species of whales. Such information will, of course, be useful to whalers, and so its publication may exacerbate the depletion of whale stocks. Information concerning technologies such as genetic engineering or behavior control through electronic stimulation of the brain allows for still more dramatic possibilities (see Tribe, 1972).

Contemporary societies are populated by the ecologically insensitive and rapacious as well as the ecologically concerned. All such interests have equal standing in the open society; thus the interest of agribusiness in high short-term crop yields, of pesticide industry workers in continued employment, or of the road transportation industry in low petroleum prices lie beyond the reach of reason therein. The only recognized power in the open society is the power of reason, but instrumental reason cannot be applied to the selection of ends.

Abandoning ourselves completely to the instrumental reason of the open society could have some dire consequences. For many supposedly instrumental choices have effects beyond any initial intended (or unintended) goals; they can also determine who or what the choosers become – and value. That is, such choices are also "constitutive" or "formative" (Tribe, 1972). Policies toward behavioral technology and genetic engineering are direct examples of such choices, but think too of the effects of a decision to (say) permit advertising on television on the degree of society's hedonism. If we give ourselves over to instrumental rationality, then we are in effect allowing ourselves to become malleable objects in an uncontrollable tide of events. The product is a society which has no influence over its future.

The content of feedback signals circulating within an open society is, then, indeterminate and arbitrary, bearing no necessary relationship to ecological values or the severity of ecological problems. Moreover, even if negative feedback concerning the condition of natural systems is present, one would expect it to be offset (to varying degrees) by other kinds of signals. Further, resilience will be achieved in the open society only if there is a widespread commitment to ecological values; but such a commitment could only be arbitrary.

The open society is wanting, then, as far as negative feedback and resilience are concerned. Unfortunately, there is worse to come. There are several reasons why coordination failure, too, is inherent in the logic of the open society.

The first such reason stems, again, from the open society's essential arbitrariness of purposes. The consequent variety of ends means that different goals can be pursued at the various decision points within society and polity (or at the same point at different times). However benign and public-spirited those goals are, the collective result in the face of non-reducible ecological problems will be failure. A problem may be "solved" in one part of the system through aggravation of a problem at some other location. "Technological fixes" are peculiarly prone to failure of this sort. For example, DDT can be used to eradicate malaria; but DDT accumulates in the environment. The consequence has often been ecosystem instability as DDT induces differential mortality rates in different kinds of organisms (not just mosquitos). Further possible ramifications include the spread of other kinds of human diseases – possibly leading to suggestions for their eradication through yet more "fixes." Operating here, of course, is the tendency toward problem displacement described in chapter 2.

Any such state of affairs is entirely consistent with the idea of "piecemeal social engineering." Processes of conjecture and refutation (free evaluation and criticism) can proceed only in a *ceteris paribus* context. That is, clear inferences about the effects of actions are possible only if one factor (or manageable cluster of factors) is manipulated at a time. If one attempts to change everything at once – a procedure usually dismissed by Popperians as "holism" – then one can make no inferences about causality. The piecemeal social engineer is very conscious of her limitations, and accepts the likelihood of numerous secondary consequences of actions. Knowledge in the open society (or, for that matter, anywhere else) cannot, according to Popperians, be centralized in the manner that holistic planning demands (Popper, 1961: 90). Therefore social problems must be attacked piecemeal. So, for example, the "layers of overlapping loops of information and control" in James's (1980) Popperian vision are at their broadest issue area-specific, at their narrowest problem-specific. In this spirit, one might choose to address (say) safety issues in the siting of nuclear power plants in isolation.

The difficulty here is that piecemeal action demands decomposition of problems; and decomposition can work only under conditions of *moderate* complexity and an absence of emergent properties in the problem area at hand. But, as discussed at length in the earlier chapters of this study, complexity and non-reducibility in the ecological circumstances of social choice mean that defensible decompositions are often hard to come by. Application of open society ideas to domains featuring complex, self-regulating phenomena such as social systems and eco-systems may be unwarranted. So, for example, safety issues in the siting of nuclear power stations overlap with questions pertaining to energy alternatives more broadly, national security considerations, proliferation of nuclear explosives, and even the degree of centralization of social, economic, and political structure (see Lovins, 1977). The arbitrary decompositions one suspects would have to characterize the open society (given a lack of agreed-upon values to use as a guide for decomposition) would stand very little chance of proving adequate. Solutions to problems defined so narrowly will be worthless if their secondary effects make matters worse overall.

Consider another example. Scotland, in common with much of Western Europe, has an acid rain problem. Fish stocks in rivers and lochs in the south-west Highlands are endangered. Other things being equal, the more forested a watershed, the more acid will end up in surface waters. The piecemeal problem-solver might respond by suggesting that the problem be ameliorated by cutting down trees, dismissing any attempt to curb acid emissions in industrial Scotland as smacking of holism. Deforestation would, of course, have ecological and social ramifications – but the piecemeal engineer would deal with such effects as they arose. Lest I be accused of contriving this example, proposals for deforestation as a solution to the acid deposition problem in the south-west Highlands have in fact been made.[9]

Decomposition itself is, of course, thoroughly consistent with the "analytic" sensibilities of Western science, which itself forms the model for the open society. Hence the circumstances inhibiting coordination – instrumental rationality, arbitrariness of purpose, piecemeal intervention, and problem decomposition – form a coherent cluster at the conceptual core of the open society.

BEYOND THE OPEN SOCIETY

The negative judgment ultimately received by the open society underscores – and helps explain – the ecological failure of the forms of social choice discussed in part II. The lack of resolution of an issue like acid rain by any type of social choice mechanism in any society (or transnational community) – a situation noted in chapter 1 – is now more comprehensible. The open society and the form of reason it epitomizes have an inherent tendency to displace problems. The open society may be the paragon of contemporary social choice, but in an important sense it really offers just "more of the same": analytical and instrumental problem-solving, arbitrary purpose, and piecemeal intervention.

[9] See the *Glasgow Herald* (August 24, 1982).

Open society virtues are substantial, and the open society concept has considerable critical utility in demolishing the claims to rationality of centralized problem-solving in the political economy. These critical claims have not been refuted here. But the redeeming features of the open society are not enough to render it ecologically attractive. Therefore a new agenda for innovation in social choice – one that goes beyond the open society – is required. Specifically, what should be sought at this juncture is a decentralized structure which does not treat norms as arbitrary (but at the same time does not seek any enforcement of norms from the center), and which allows for the exercise of forms of reason more conducive to the symbiotic intelligence of ecological rationality than is the instrumental rationality of the open society. A structure of this kind will be sketched in the next chapter.

15

Practical Reason

The problem-solving attractiveness of the open society stems largely from its promotion of reason. Reason in the open society is instrumental: an actor or institution is rational to the extent of its capacity to select and effect actions in successful pursuit of some end. Moreover, this conception of rationality is prized in, and reinforced by, the dominant forms of social choice in contemporary society. And many would-be reformers of these mechanisms are themselves generally under the sway of an instrumental conception of rationality.

The previous chapter demonstrated the inadequacy of purely instrumental and analytical problem-solving sensibilities in an ecological context. This chapter will articulate a kind of social choice inspired by a type of rationality ignored in the open society: "practical reason." My intent is to demonstrate that there is more to practical reason than seemingly abstruse philosophy, and to work toward concrete proposals for social choice. The resulting model departs considerably from the open society (and still further from the dominant kinds of social choice in today's world). I will claim that this model promises enhanced ecological rationality in collective choice.

This study's exploration of social institutions is now at a transition from instrumental rationality to practical reason. If the claims of practical reason hold up, this transition might profitably presage a turning point in human affairs more generally.

COMMUNICATIVE RATIONALIZATION

In the spirit of Aristotle, with whom the term "practical reason" is most strongly associated, social choice involves a collective cultivation of virtuous behavior, rather than the administration or manipulation of people and things. Aristotle's contemporary heirs include nostalgic republicans such as Arendt (1958), participatory democrats such as

Barber (1984), and critics of post-Enlightenment moral relativism such as MacIntyre (1981). But the flame of practical reason today burns most strongly in the field of critical theory, whose language will therefore be used to develop the argument of this chapter.

Critical theory's counterpart to the idealized open society is encapsulated by Habermas in what he terms an "ideal speech situation" (Habermas, 1973a). Discussion in this situation is thoroughly unconstrained; there are no restrictions on who may participate, or on what kinds of arguments may be advanced, or on the length of deliberations. The only resource available to participants is argument, and the only authority is that of the better argument. It is crucial, too, that all participants possess roughly equal degrees of "communicative competence" (Habermas, 1970b), the capacity to participate effectively in discussion by making and challenging arguments. Under these conditions, collective choices can be arrived at only discursively, and the product is an action-oriented consensus.

Unlike the situation obtaining in the open society ideal, collective actions in conditions of free and open practical reasoning will not necessarily be chosen because they are good means to a given (and arbitrary) end. As Habermas (1973b: 42) puts it, political deliberation should proceed "pedagogically" rather than "technically." Thus, the likely course of action is the one supported by the best reasons, reflectively generated by the group. "Good reasons" here *can* be instrumental reasons, but they can be much more besides. In particular, good reasons can refer to the moral rightness or wrongness of an act, or of the goals toward which that act seems to be directed, or of the kind of society which that act would help constitute, or (to use the Aristotelian language) the virtues with which the act is consistent. Hence actors can rationally question the normative grounds for the positions held by other participants – for example, by holding them up to shared value systems, or through reference to their consistency with norms embedded in a shared culture, or by elaborating the broader consequences of universal adherence to such a normative stance.[1]

Clearly, the precepts of ideal speech may not be located in any existing form of social choice. Polyarchy is perhaps closest, though still far from the ideal. Polyarchy is subject to the "systematically distorted communication" (Habermas, 1970a) which can result from the exercise of

---

[1] There is some similarity with legal argumentation here; indeed, in the context of policy analysis, this kind of process is sometimes referred to as "forensic social science" (Rivlin, 1973). However, legal argumentation is adversarial and aggressive; the ideal speech situation allows disagreement, but its spirit is cooperative and reflective (cf. Barber, 1984: 175).

political power, the strategic use of language to manipulate and deceive on behalf of particular interests, ideology and dogmatism, the debasement of everyday language through advertising, the debasement of political language by professional politicans and their consultants, and so forth. Any consensus attained under such conditions will be severely "distorted." At worst, that consensus will merely reflect the interests of the political powers that be. Even at best, though, such consensus will constitute only a compromise between competing special interests.

Any consensus influenced by the modes of distortion noted above is less defensible for that admission (McCarthy, 1978: 309). Conversely, the closer an actual choice context approximates the ideal speech situation, the more defensible is its outcome. Hence the real value of the ideal speech situation is as a *standard* which real-world discussions will approximate to greater or lesser degree. Like the open society, the counterfactual state of affairs posited in the ideal speech situation is not a blueprint to be fully achieved in practice, but it can be used to evaluate actual, proposed, or conjectured social arrangements.[2] In particular, any practices, institutions, or decisions which could be justified only by departure from the precepts of ideal speech merit clear condemnation. More positively, the potential for shifting terms of discourse toward ideal speech principles can be explored. Habermas (1984) refers to this kind of development as "communicative rationalization" of social and political life. It is the prospects for and ecological implications of such movement which will be investigated here.

Given the "openness" and decentralization they share, to what extent is the communicative rationalization of critical theory an improvement over the open society?

The open society deals only in the exercise of instrumental problem-solving (see chapter 14); purpose is arbitrary. In contrast, the ideal speech situation can be applied to discourse about morality as well as about truth. Norms themselves can be discursively generated and validated. Any individual (or other actor) contemplating action based on a normative principle can be called to account for that principle: "I must subject my maxim to all others for purposes of discursively testing its claims to universality" (McCarthy, 1978: 326).[3]

---

[2] The scientific community on which Popper models the open society is, in reality, characterized by varying degrees of authority, discipline, and punishment of transgression, not to mention fad and fashion. Some historians and philosophers of science construct accounts of the enterprise of scientific inquiry based on such extra-rational elements (for example, Kuhn, 1962).

[3] As McCarthy notes, this testing represents a "procedural reinterpretation of Kant's categorical imperative" (the idea that moral individuals should adopt a maxim only if they would wish it to become a universal law).

This notion of discursive generation and validation of moral judgments enables Habermas (1973a) to distinguish between idiosyncratic or
selfish "particular" interests, and "generalizable" interests, common to
all individuals and attainable through rational discussion. Particular
interests cannot prevail in the perfectly unconstrained discussion of the
ideal speech situation, for widespread (still less universal) agreement is
impossible about them. Recognizing these conditions of public debate,
any citizen would come to regard it as improper to even advance a
particular claim.

Practical reason provides for the proposal, development, and rational
acceptance of common interests, purposes, and values – in direct contrast
to the open society.[4] It should be stressed, though, that communicative
rationalization of social interaction does not guarantee agreement on
norms. Participants may continue to disagree, even after the substantial
reflection which encounter with others holding different positions forces
upon them. Disagreement may persist as a result of different life
experiences, or incompatible conceptions of human nature. Even so,
consensus rooted in reasoned *dis*agreement is still possible. Participants
can agree upon the "what" of action without full subscription to a
common normative judgment. Here, full comprehension of the reasons
why others hold to different normative positions is crucial if the resulting
consensus is to be described as rational. Thus it is not different particular
interests which persist, but different conceptions of generalizable interests. One can therefore expect convergence on a smaller number of
norms, if not perfect normative agreement.

The decision rule of consensus can therefore be preserved in the
absence of bland uniformity on norms, though particular interests have
to fall by the wayside. Moreover, the content of consensus can be
adjusted in the light of experience. Practical reason does not involve a
search for eternal verities (cf. Barber, 1984: 170), or any mystical and
infallible "general will" of the sort proposed by Rousseau (1968).[5]

There is more to the action-orientated consensus of practical reason
than mere arid theorizing. In their very practically minded analysis of
effective negotiating techniques, Fisher and Ury (1981: 84–98) recognize
that agreement between negotiating principals is facilitated if it is sought

[4] This is a slight exaggeration, as the open society implicitly demands universal
acceptance of the canons of reason.
[5] While Rousseau believes in participatory democracy, at least in small communities
faced with simple issues, it is not clear that he believes the general will can be attained
through the kind of reasoning process described here. Indeed, Rousseau has such little faith
in the ability of a community to determine its own best interest that he introduces a
"lawgiver" to save it.

through reference to what they term an "objective criterion," independent of the particular interest or position of each participant. According to Fisher and Ury, this result holds because one can reason about – and with reference to – an "objective criterion" (be it fair market value, or average comparable settlement, or parity), whereas it is possible only to fight about (particular) self-interest.

ECOLOGY AND PRACTICAL REASON

If the content of norms and action principles generated by communicatively rationalized political debate were sensitive to ecological values, then clearly this would impose a content to negative feedback signals and collective problem-solving efforts that is lacking under the open society's institutional arrangements. But would ecological values come to the fore in reality?

In most ecological problems, the actors involved have both particular and generalizable interests. Particular interests might include the use of a river for the disposal of one's waste products, avoidance of any contribution of one's personal resources to the clean-up of a toxic waste dump, or maximization of one's take from a common property resource such as a fishery. The corresponding generalizable interests would be the quality of the river's water, the cleanup of the dump, and sustainable total yield from the fishery. Clearly, any form of social choice promoting generalizable over particular interests is going to stand ecological values in good stead.

While one could not be certain that ecological values would be primary – for example, people might make a considered choice to exploit resources to the limit, and thoroughly discount the future – the fact that the integrity of the natural systems upon which human life depends is an obvious generalizable interest places ecological values in a strong position. Public discussion of the moral desirability of energy conservation in the wake of the energy crises of the 1970s is indicative of the possibilities here. Moreover, as I have argued in chapter 5, the human life-support capacity of natural systems is *the* generalizable interest *par excellence,* standing as it does in logical antecedence to competing normative principles such as utility maximization or right protection.[6]

The content of any norms reached through communicatively

[6] Of course, I cannot guarantee that individuals in communicatively rationalized situations would actually agree on the life-support principle. They might come up with something better.

rationalized discourse is not invariant. Rather, those norms can be contingent on time and place. This possibility accords with temporal and spatial variability in the ecological circumstances of social choice (see chapter 3). For example, a community norm that encourages the installation of wood-burning stoves may achieve generalizable status inasmuch as it reduces dependence on scarce and unreliable supplies of non-renewable energy; but that norm may stand adjustment if so many people convert to wood-burning that local deforestation and extensive wood smoke pollution result.

Once the content of social norms or broad principles for action have been agreed upon, then the creative energies of individuals, in isolation or in combination, could be released to solve problems. Communicative rationalization promotes the cooperation of individuals concerned with different facets of complex problems. Here, practical reasoning is less cumbersome than bargaining, for only norms and broad principles are "negotiated," not the detailed contents of actions. This creative spirit is reminiscent of that obtaining under moral persuasion (see chapter 11). The difference, of course, is that moral persuasion imposes norms from above, and hence is more likely to face a compliance problem. The fact that actors have freely consented to norms or principles for action enhances the likelihood of subsequent compliance. This pattern is noticeable in some American cases where consensual dispute resolution techniques have been applied to site-specific environmental conflicts (see Mernitz, 1980: ch. 4).

The likelihood that ecological concerns will be reflected in social norms in communicatively rationalized settings could be enhanced, one suspects, if the community in question were small-scale and self-sufficient. Smallness of size facilitates consensus in discursive value formation, and autonomy lends an immediacy to ecological concerns that would promote their acceptance as rational norms. This question of autonomy and scale will be explored at length in the next chapter; for the moment, the question remains an open one.

The indeterminacy in content of negative feedback signals and collective problem-solving efforts resulting from the value-arbitrariness of the open society has now been dispensed with. But no escape has yet been engineered from the instrumental and manipulative orientation toward the *natural* world which the open society fosters. Does practical reason offer or promise any improvement over the analytic, piecemeal, and instrumental approach to problem-solving of the open society?

Certainly, the sensibilities of a community of participants in any communicatively rationalized situation are not amenable to grafting onto the instrumental problem-solving of the open society. Under the communicative rationality ideal, there is no distinction between human subjects and human objects ("target population," in the language of social policy analysis). Nor can any distinction be drawn between "treatment" and "control" groups in social experimentation.[7]

What can replace the piecemeal experimentation – and tunnel vision – of the open society? On the assumption that imperfect knowledge requires *some* form of exploratory trial and test, the only real alternative is some kind of holistic experimentation. In the holistic case, the external validity of experimental results – that is, generalization to the universe of which the "treatment" group is a subset – is irrelevant, for no such universe exists outside the system to which the "treatment" is being applied. But even to speak of "treatment" is misleading, for there can be no division between experimenters and their targets. Instead, all individuals can participate in the design of the experiment, human subjects in the true sense of the term. "Internal" success of the experiment is all that matters, promoted by constant innovation and enthusiasm on the part of participants. Holistic experiments are, then, dialectical rather than instrumental in spirit (Mitroff and Blankenship, 1973).[8]

Holistic experimentation of this sort is a practice that promises escape from the rationalistic dogma of the open society; but can it enable escape from purely instrumental human orientations toward the *natural* world?

The formulations of mainstream critical theorists and participatory democrats are of little assistance here. For example, Habermas sees technical or instrumental knowledge – natural science – and manipulative forms of practice as thoroughly appropriate to human dealings with the natural world. Practical reason, manifested most strongly in what Habermas refers to as "emancipatory" knowledge, is reserved for the social sciences and social practice. Thus Habermas sees a discontinuity between the systems of the human world (potential subjects) and those of the natural world (inevitable objects). From the viewpoint of ecological rationality, this discontinuity is a misplaced decomposition of a non-reducible system.

[7] Control group experimentation is the method most favored by piecemeal social engineers; the conditions of a treatment group are manipulated (for example, through the application of some kind of prototype public policy), and its subsequent performance is compared with that of an otherwise identical control group (see Campbell and Stanley, 1966).

[8] Holistic experimentation as outlined here differs from Popper's characterization of the process. In his demolition of holism, Popper identifies holistic experimentation as a *centrally planned* kind of social change (Popper, 1961: 83–93). Clearly, a dialectical and participatory spirit is totally inconsistent with central planning.

A charitable interpretation of critical theory's position here would regard it as an oversight; critical theorists (with the possible exception of Horkheimer, 1947) have traditionally had little knowledge of (or concern for) ecology. Therefore they may be unaware that ecology is not a predictive science, only an explanatory one. Ecosystems are not amenable to instrumental manipulation because they are non-deterministic, so that one cannot forecast the results of any novel human intervention. An extensive range of consequences can result from innovative actions (such as the release of exotic predatory insects into an agro-ecosystem for pest control purposes).

On the other hand, critical theory's attitude toward the instrumental manipulation of ecosystems may be rooted in a recognition that to be accorded full subject status an entity must have the potential to participate in social discourse. Clearly, the entities of the natural world fail this test.

However, while ecosystems cannot literally "speak" to human subjects, they can communicate in other ways; especially, of course, through feedback signals (which can be mapped onto social discourse). If the topsoil on which my crops depend is shrinking, then clearly nature is "telling" me something. Moreover, though they are not teleological, ecosystems do possess spontaneous self-organizing and self-regulating qualities. Ecosystems do, then, have some tentative claims to "subject" status. But what, concretely, is the implication here for social practice and social choice?

The implication is this. Symbiotic practice is forever beyond the reach of the instrumental sensibilities of the open society, under which nature *can only* be manipulated and engineered. While there is no guarantee that any more symbiotic orientation toward the natural world would prevail under communicative rationality, there is no obstacle to such an orientation therein. Indeed, human systems and natural systems in combination clearly constitute complex non-reducible systems amenable to holistic experimentation of the sort discussed above. The human participants in any such non-reducible system can choose to treat the natural elements *as if* they were subjects – that is, treat ecosystems as teleonomical (as opposed to teleological) entities. Therefore human subjects would both allow for and react to the spontaneous "actions" of ecosystems. Ecosystems do, in fact, have stable "goals," even if nobody or nothing consciously sets them (Patten and Odum, 1981: 888). As a practical matter, one could even choose to appoint spokespersons for ecosystems – in a manner analogous to Stone's (1972) suggestion that human "guardians" be appointed to represent the legal rights of natural objects. It should be stressed that such treatment of ecological systems would be justified here in terms of human interests (by the ecological rationality standard), though some philosophers have argued for similar treatment through reference to nature's "interest."

Symbiotic practice of the sort discussed here is thoroughly consistent with one critical theorist's conception of reason. Max Horkheimer contrasts the "subjective" (instrumental) reason of industrial man with an older "objective" reason (adhered to by Plato and Aristotle, among many others). Objective reason "aimed at evolving a comprehensive system, or hierarchy, of all beings, including man and his aims. The degree of reasonableness of a man's life could be determined according to its harmony with this totality" (Horkheimer, 1947: 4).

There currently exist few examples of self-conscious holistic experimentation to shed light on the capabilities of this form of collective choice.[9] There exist, to my knowledge, none at all that attempt to apprehend ecosystems in the manner I have suggested. Let me attempt to remedy this empirical dearth by sketching a hypothetical case – admittedly highly contrived, but dealing with a pressing set of ecological problems. My example is based on the so-called "green revolution."

The intent of the sponsors of the green revolution was to alleviate hunger in the Third World by increasing crop yields. The mechanism through which this goal was to be accomplished was the introduction of new strains of "miracle wheat" and other cultigens. The green revolution is an excellent example of instrumental human intervention in (agro-)ecosystems. In terms of its primary objective – increased yields per acre – the program was a spectacular success in a number of locations, especially the Punjab in India.

Unfortunately, this apparent success has a price. The "miracle crops" require large inputs of water – necessitating irrigation – and nitrogen fertilizers. Irrigation leads to the progressive accumulation of salt in the soil. Nitrogen fertilizers destroy the natural nitrogen-fixing agents in the soil, and ever-increasing quantities must be applied to keep yields constant. Moreover, these crops are not very resistant to pests and weeds, so they require extensive application of synthetic pesticides and herbicides. A technological race against the genetic adaptive capacities of pests is thereby set in motion. The harvesting of miracle crops is most efficient using high-technology machinery; therefore small landholdings become uneconomical, and a rural landless class grows.

The green revolution's replacement of traditional agro-ecosystems by unstable monocultures is, then, capable of promoting ecological and

[9] For that matter, there exist precious few good examples of systematic piecemeal experimentation, either.

social catastrophe – not because of its failure, but because of its success (at least in terms of its stated goals).

One might argue that this result stems simply from a failure to anticipate the diverse "secondary" effects of an intervention. Can one not plan for such effects before the fact – or, alternatively, cope with them as they arise through further instrumental manipulation?

Unfortunately, it may be that each instrumental action leads to a net *increase* in the overall severity of problems (cf. Ehrenfeld, 1978: 107–12). Consider, for example, the number of ramifications of a superficially straightforward action such as the introduction of a new strain of wheat – just a few of which were mentioned above. The dynamics of instrumental rationality in an ecological context may lead to a proliferation of problems rather than convergence on any "less problematical" state, even if, in terms of its own intended goals, each instrumental action is a success. A piecemeal approach to problem-solving inevitably suffers from tunnel vision. Moreover, it is no co-incidence that, in the green revolution case, each secondary problem elicits a certain kind of response. Once miracle crops are in place, the *only* available response to (say) disease or pests will be a synthetic one. So the supposedly piecemeal first step – the new crop – is effectively the first move on a slippery slope of logically connected manipulations.

How might the green revolution have looked had it been conducted as a holistic experiment? The boundaries of the experiment would have been set by particular agro-ecosystems – indeed, it would make no sense to think in terms of *one* experiment (directed from Rockefeller Foundation headquarters in New York), but only of many different experiments. Uttar Pradesh presents a different range of problems and potentialities than does the Punjab.

Having defined the bounds of each experiment, the participant members of both human systems and natural systems could be identified. Indeed, these human subjects could assist in the definition (or re-definition) of the bounds of the experiment, who or what to include, and the goals toward which the experiment is to be directed. Any "experimenter" need not take the human subjects as they are, but could perhaps educate them and otherwise promote their "communicative competence" – and, further, be prepared to learn from these subjects. Having thoroughly blurred any distinctions between experimenter and "target population," the next step would be to identify problems. One hopes that problems would be defined broadly, in terms of sustainable life support from the ecosystem in question.

Little can be said in advance about the specifics of the actions that might be chosen in any such holistic experiment. One suspects, though, that a request to New York for miracle rice would be highly unlikely. A concern for the whole system – ecological, social, and cultural – would lead to modesty in actions taken. No "revolution," green or otherwise, would occur. Any actions taken, then, could be sensitive to local ecological conditions, cultural norms, and traditional agricultural practices. Imbued with a sense of the whole, individuals and groups would be free to innovate as they saw fit – for example, with intercropping, or new combinations of crops, or soil conservation practices, or community-controlled reforestation that used indigenous species of trees, or even land reform. Individuals would not be bound by the piecemeal experimenter's need to keep "treatment" constant over the experimental period. For ecosystems, like social systems, are dynamic and self-organizing; therefore there is no static *ceteris paribus* against which interventions can be judged. Nor need individuals be restricted to actions based on "objective" or scientifically testable knowledge – instead, they would be free to make use of tacit knowledge about the systems with which they were familiar. Practical reason places no restrictions on the kinds of knowledge admissible. However, individuals would be subject to the action principles agreed upon by the community, which might rule out piecemeal actions such as those of a single farmer wanting to use a synthetic pesticide on his crops. The fact that all individuals have a say in decisions means that each proposed intervention would be tested in advance – and evaluated in retrospect – against a wide variety of concerns. This breadth of evaluation could act as an antidote to the tunnel vision of instrumental rationality. But conscious effort would still need to be made to monitor effects also on whole systems, both human and natural.

Certainly, one could not guarantee that the results of any such holistic approach would be positive. Yet at least this approach would allow for the possibility of practical reason in terms of both a shared commitment to ecological principles and symbiotic orientation to the natural world, whereas instrumental interventions of the green revolution variety do not. I do not mean to paint too rosy a picture; there are considerable rigidities in contemporary Indian rural society that preclude any extensive reform of agricultural and social practices in the direction indicated. On the other side of the coin, the green revolution is not the only source of India's failure to provide sustainable life support for its population: this failure predates the green revolution.

It is also worth noting at this point that practical reason facilitates coordination in social choice. The participants in communicatively rationalized processes can generate shared norms and intersubjective understandings more generally, and therefore can act in concert in ways precluded by the piecemeal approach to action, value-arbitrariness, and responsiveness to "particular" interests of the open society. Moreover, the discursive processes central to communicative rationalization are likely to promote "cooperative" over "defecting" strategies in the prisoner's dilemma. One persistent finding of the experimental literature on the prisoner's dilemma is that a period of group discussion prior to each individual making a choice on how to behave sharply increases the frequency of cooperative behavior within the group (see, for example, Jerdee and Rosen, 1974; Dawes, McTavish, and Shaklee, 1977). Thus practical reason promises substantial advance on the "coordination" aspect of ecological rationality.

DESIGN FOR PRACTICAL REASON

The analysis of this chapter to date has proceeded at a somewhat abstract level. How, concretely, may *institutions* (as opposed to practices of the sort discussed in the green revolution example) for the promotion of practical reason in social choice be identified or designed?

In his presentation of the ideal problem-solving community, Popper has the advantage of some existing systems – liberal polyarchies and scientific communities – to offer as rough approximations to the open society. In contrast, communicative rationalization has no such obvious empirical referents. Any specification of communicatively rationalized social choice is consequently difficult.

The formulations of critical theorists themselves are of little assistance here. While recognizing the need for communicatively rationalized institutions (see, for example, Wellmer, 1985: 58), they tend to use extremely abstract language, and retreat to obscure generalizations when presenting ideas for real-world social institutions. Critical theory can be a fairly obscure enterprise. As yet, its proponents have said little about institutions which would promote "discursive will-formation" in any set of concrete circumstances.

If one cannot deduce any implications for institutional structure, however modest, from the idea of communicative rationalization, then one might as well reject it. The challenge is to cast aside the cloak of ambiguity in which discussions of the practical implications of communicative rationalization are typically shrouded.[10] There is no need to

---

[10]This ambiguity may result from the uncertain position of critical theorists with regard to revolutionary change. Critical theory has a revolutionary heritage, but its logic is thoroughly reformist. Nevertheless, Habermas himself continues to believe that unconstrained communication is only possible once exploitative material conditions have been abolished (see Bernstein: 197).

articulate Utopias. The point is to devise institutions which will "justify the presumption that basic political decisions would meet with the agreement of all those affected by them if they were able to participate without restriction in discursive will-formation" (McCarthy, 1978: 332). Such institutions would have to be sensitive to social and cultural variety, and the constraints and opportunities of particular situations. There are, in fact, some straws in the wind in the real world of social institutions.

In a world of strongly held "particular" interests, political machination, linguistic manipulation, and overt exercise of power, an important precondition to communicative rationalization is the establishment of concern with social values over and above the particular strategic interests of the individuals involved.[11] The establishment of such an identity may require in its turn the auspices of a "third party" to set up a discursive forum, and to introduce participants to an unfamiliar style of interaction. That third party can be neutral with respect to the issues at hand.[12] Alternatively, the third party function can be performed by all the participants agreeing to abide by a set of rules (which can be introduced from within or without). Having set the proceedings in motion, any third party can then fade into the background.

Some of the more successful arrangements of this sort have taken the form of *mediation*. Traditionally, mediation has made most of its appearances in labor disputes and international crises. More recently, though, mediation of environmental disputes has been attempted in a number of cases, especially in the USA. Mediation is a process of collective choice in which the interested parties – normally involved, at least at the outset, in some kind of dispute – agree to discuss and reason through their differences under the supervision of a third party. Unlike a judge or arbitrator, that third party does not produce binding decisions – or, indeed, any decisions at all. The mediator can, though, play an active role in the specification of normative judgments, novel definitions of the problem at hand, and proposals for action (see, for example, Wall, 1981).

In a forum of this kind, possibilities exist for the more or less formal articulation of rules of discourse, such as those outlined in Fisher and Ury's (1981) "principled negotiation": a conscious separation of individual egos from joint problem resolution, a stress on interests rather than negotiating positions, serious efforts to devise actions involving mutual gain, and the determination of solutions through reference to

---

[11] For a discussion of this requirement in an organizational context, see Thayer (1981: 29–34). More generally, see Riddick (1971).

[12] Though, as Touval (1982) notes, biased third parties can be useful too.

"objective criteria," independent of any particular interests. These rules can themselves be modified, though, through the consent of the parties involved – the mediation process generally involves large elements of improvisation (see Clark, 1980).

Any outcome of mediation is a product of consensus among the participants. At worst, that consensus may represent co-optation of the relatively powerless (for example, community groups and environmentalists) by the relatively powerful (for example, corporate developers). Thus one might get a veneer of "responsible development," but the shopping mall or nuclear power station will still be constructed (Amy, 1983). Only slightly less discouraging are cases where consensus is a mere compromise between competing particular interests, sufficient to settle the conflict in the issue at hand. Mediation as dispute resolution does no more than lubricate polyarchy or a legal system. If dispute settlement is the aim, then normative discourse is best suppressed; environmental mediators in the USA generally seem to believe that fundamental "ideological" differences can be an insurmountable obstacle to mediation, and hence are best avoided (for example, Talbot, 1983: 93). In a polyarchical context, it may indeed be possible to get actors to agree on a policy, without agreement on the reasons for it (Lindblom, 1959: 83–4).

Compromise and conflict settlement as such are, of course, not particularly interesting from the perspective of ecological rationality (even though they are likely to promote social welfare in economic efficiency terms). It is entirely possible for an agreement thus reached to allow for continued systematic destruction of ecosystems. For example, fisheries quotas which when summed exceed sustainable yield might be agreed upon – note the continued depletion of whale stocks despite quotas negotiated in the International Whaling Commission (see Dryzek and Hunter, 1987). In this context, it is particularly unfortunate that most proponents and practitioners of environmental mediation have stressed the value of dispute settlement *per se,* and have studiously ignored the *content* of mediated accords (see, for example, Mernitz, 1980). Thus one observer considers mediation of a dispute over power plants on the Hudson River a success (Talbot, 1983: 7–24) – even though the agreement reached allows for the discharge of large quantities of warm water into the Hudson, and the facilitation of highly centralized energy generation (which distresses the environmental sensibilities of, among others, Lovins, 1977).

What this attitude means is that the real promise of environmental mediation – the promotion of practical reason – often goes unfulfilled at present. Indeed, many proponents of mediation suggest that it should have no role in cases (such as nuclear power plants) in which there are deep differences in the positions of the concerned actors, on the grounds that principled positions (for example, either for or against nuclear energy) are not amenable to compromise (see, for example, Cormick, 1976: 218).

Despite such attitudes, mediation at its strongest can produce coherent strategies through discursive processes such that generalizable interests do indeed come to the fore. For in a mediation forum it is – ideally – the power of reason that holds sway: the normal tools of political conflict (such as manipulation, formal authority, or propaganda) are discarded. If it is to be persuasive, any argument offered in that forum cannot be based on the particular interest of any party (or parties) involved. Any persuasive argument must, instead, be generalizable to all the parties. Norms and values can be – and are – subjects of discussion in a mediation forum, and the values of participants can and do change as a result of participation.[13] Indeed, it may be that staying at the level of positions or particular interests will produce only a stalemate (as Fisher and Ury, 1981, recognize). A shift to generalizable interests may be the only way a "contract zone" can be created. And, as already noted, a focus on generalizable interests stands ecological concerns in good stead.

If one judges success in terms of consensus among initially hostile actors, then the record of environmental mediation in the USA is quite positive. Mediation has been applied to cases as varied as New Jersey's coastal zone management, New York City's Westway highway, California's power plant siting, flood control in Washington State, uranium mining and milling in Colorado, and coal conversion in New England (see Lake, 1980, and Talbot, 1983, for surveys of cases). Nearly all of these cases have been site-specific disputes among organized collective actors, such as developers, environmental groups, community associations, and local governments. One should, of course, be wary of interpreting agreement as proof of success; but clearly, environmental mediation is a feasible form of social choice.

Environmental mediation as an intimation of communicative rationalization is currently hindered because collective actors can only send representatives to a forum. Hence the outcome of any process of discursive will-formation, while common to the individuals actually participating, may exclude and perhaps alienate the people who do not

---

[13] So, for example, in the mediation of a dispute over the construction of an interstate highway in Seattle one participant explains a change in position in the following terms: "We bent on the transit-lane issue because I couldn't see how it was right for people in 1976 to tell future users how each and every lane was going to be used" (Talbot, 1983: 37).

participate. What this recognition suggests is that a mediation-like process might produce the most positive results in relatively small communities, at least as far as the promotion of practical reason is concerned.

CONCLUSION

Practical reason in social choice promises some major advances according to the ecological rationality standard. In particular, this model holds out the possibility of significant improvements in negative feedback, resilience, and coordination. There are reasons to suppose that this possibility is most likely to be realized in a setting of small and relatively autonomous communities. The further ecological implications of small-scale social choice and their relation to practical reason will be sketched in the next chapter as the second item on an agenda of innovation.

16

Radical Decentralization

Both open society instrumental problem-solving and practical reason are obstructed to the extent that *hierarchy* exists. This chapter will explore the prospects for radically reducing – perhaps even eliminating – hierarchy in social organization. The resulting radically decentralized social choice will be non-hierarchical, but – equally important – it will involve local autonomy and small scale in social organization. Local autonomy provides a way of "mapping" ecological feedback signals onto social choice. Small scale can promote practical reason, for the discursive processes central to communicative rationalization are facilitated to the extent that individuals interact with identifiable others on a frequent and recurrent basis. Such frequency of interaction has the added advantage of promoting cooperative solutions to the prisoner's dilemma (as will be shown below). Therefore coordination within the interacting group is enhanced. The design sketched in this chapter can thereby kill several birds with one stone, and constitutes the second major element on an agenda of institutional innovation.

PRACTICAL ANARCHY

Any radically decentralized system may also be described as anarchical. Anarchy means, quite literally, the absence of a state, and therefore of state authority. This characterization should not be taken to imply that authority and power are totally lacking, for authority can stem from possession of a better argument (Taylor, 1982: 24), and power can be exercized on a decentralized basis. A pure anarchy, then, may be defined as a social system in which there is no concentration of force and no political specialization (Taylor, 1982: 9). Therefore there exist no *formal* institutions at the system level (cf. Young, 1978a).

This definition of anarchy accentuates the negative – what radically decentralized systems lack. That kind of definition is relatively un-controversial. It is in the determination of what anarchy *does* contain that room for divergence enters. Numerous and varied accounts of highly decentralized systems exist in the conjectures both of writers who call themselves anarchists, and of those who eschew that label. Real-world examples of highly decentralized social choice include the technical anarchy of the international system (discussed in chapter 12), preliterate communities of hunter–gatherers and horticulturists, some peasant communities,[1] communes scattered within contemporary industrial societies, and broad-scale communes established on a few brief occasions (Paris in 1871, the Ukraine in 1917, and Spain in the 1930s) by self-conscious anarchists.

Rather than scrutinize all the available options and permutations of highly decentralized systems, I will follow a more selective "design" approach here. Design will be directed toward enhanced practical reason, feedback mapping, and coordination in social choice.

WHY SMALL IS BEAUTIFUL

Innovations for the promotion of practical reason of the sort sketched in the previous chapter are one way to overcome the value-arbitrariness and concomitant indeterminacy in the content of feedback signals and collective problem-solving in the open society. In this section I will explore a second way of forcing attention to ecological feedback signals: a "small is beautiful" principle in social choice.[2]

One central feature of smallness of scale is that a locality both relies upon and has exclusive jurisdiction over the productive, protective, and waste-assimilative functions of the ecosystems in its immediate vicinity. Local self-reliance makes one of its better known appearances in E. F. Schumacher's "Buddhist economics": "production from local resources for local needs is the most rational way of economic life" (Schumacher, 1973: 59).[3] So, for example, Schumacher roundly condemns the large-

---

[1] Though subservient to lord or state, many such communities are (or were) internally anarchical (see Taylor, 1982: 35).

[2] While not all highly decentralized systems are small-scale, the desirability of smallness of scale is a recurrent theme in the writings of anarchists and others sympathetic to this tradition.

[3] The guiding principle in Buddhist economics is the maximization of wellbeing at a minimum of consumption, as opposed to the consumption maximization which informs mainstream economics (Schumacher, 1973: 57–8). The finer points of Buddhist economics need be of no major concern here.

scale monoculture of the Canadian prairies – a thoroughly rational
arrangement according to mainstream economic precepts. As he puts it
to the wheat farmers of Saskatchewan, "make up your minds whether
you want to die economically or survive uneconomically" (Schumacher,
1979: 104).[4]

Local self-reliance of this sort means, first and foremost, that
communities and their members must pay great attention to the life-
support capacities of the ecosystem(s) upon which they rely. This
attention is not a matter of choice, other than choice between life and
death. Self-reliance bars access to – and exploitation of – distant
"invisible acres" (Catton, 1980), and rules out despoliation followed
by emigration. Local residents are forced to heed negative feedback
signals from their natural environment. Such signals have an immediacy
and clarity which is generally lacking for most members of contemporary
industrial societies (however "open"), who have scant knowledge of – let
alone concern for – the ecosystems upon which they depend. Consumers
of fast-food hamburgers in the USA are generally unaware of their role in
depleting tropical rain forests in Central America.[5] One would expect a
community depending on a rain forest ecosystem for its survival to
exhibit greater concern for the forest than do Burger King customers,
even though the latter may eventually come to feel the effects of
destruction of the world's ecosystems.

Self-reliance shortens feedback channels, and so ecological signals are
less diffuse and more readily "mapped" onto social choice. There is little
scope for the obstruction and distortion of such signals. So, for example,
a community of Inuit whalers will probably be in a better decision to
make decisions regarding the feasibility and environmental effects of a
sustainable-yield harvest of marine mammals in its vicinity than members
of a federal agency located in Washington DC or Ottawa.

The experience of small-scale, self-sufficient human societies – whether
internally anarchical or merely autarchical – suggests that isolated and
largely autonomous communities do indeed exhibit a high sensitivity and
responsivenes to feedback signals from the environment (see, for example, Bernstein, 1981). Indeed, it could hardly be otherwise, for autonomous communities which did not exhibit such traits – or develop them
when under stress – would surely perish in a process akin to natural
selection. Thus ecological anthropologists interpret social and economic

---

[4] Other well-known works with similar emphasis on self-reliance are those of Goldsmith
et al. (1972), Bookchin (1980, 1982), and Sale (1980).
[5] A major cause of deforestation is the expansion of cattle pasture for beef exports (see
Guess, 1979).

structures and individual beliefs, values, and behavior patterns as adaptive devices for coping with stresses in the human or natural environment (see chapter 4).

Other things being equal, then, the integrity of ecosystems is highly visible to members of small-scale, self-sufficient communities. However, ecological resources themselves vary in their visibility. So, for example, simple ecosystems are more readily comprehended by their human populations than are complex ecosystems; and the depletion of a species that occupies a fixed habitat is ascertained more readily than that of migratory species, as Nelson (1982) notes in the case of the Koyukon of Alaska. Therefore a range of degrees of sensitivity to feedback signals can be expected in otherwise similar autonomous communities.

Preliterate anarchies, apart from being small-scale and largely self-reliant, possess an additional attraction in terms of the content of feedback signals and collective motivation. For many such societies had – and have – orientations toward the natural world which may be described as "organic" as opposed to instrumental in sensibility. It is in this vein that many anarchists look to preliterate communities for elements of a model for future progress (for example, Bookchin, 1980, 1982). The anarchist tradition sees nature in terms of a "great chain of being," of which preliterate man was a part, but from which agricultural and industrial man has temporarily escaped, only to inflict disasters upon himself (see Woodcock, 1977: 17). There is no hierarchy of superior and inferior species in this great chain; therefore Bookchin (1982: 25) can argue that an understanding of ecology can underpin a non-hierarchical ontology, or conception of reality – a "oneness" of living things. Without hierarchy, there is little scope for instrumental manipulation of man by man, or of nature by man.

This ontology underpins a replacement of the instrumental rationality of domination by the symbiotic practice of cooperation. Even human predation upon animals can be interpreted in this light. So, for example, the Inupiat of Alaska ascribe a spirit and intelligence to the Bowhead whales they hunt which are, if anything, of higher status than their own. The practice of cooperation in this kind of worldview is consistent with co-evolution in the relationships between species, which work to benefit the ecological community upon which the population of any one species depends.

A recognition of these ecologically rational aspects of traditional societies such as preliterate anarchies – enhanced feedback sensitivity through an "organic" sensibility – does not, though, warrant a recommendation that contemporary industrial or agricultural societies simply return to that form of social organization. For clearly, many of these societies achieved harmony with their environment on a largely unselfconscious basis (see Alexander, 1964: 46–54). So, for example, Owen (1973) notes that peasant farmers in tropical Africa are often hesitant to remove weeds or pests from their crops, even when it requires very little effort. The farmers themselves cannot explain their behavior, but Owen discerns good ecological reasons for this inaction. (For example, weeds may be good hosts for predators of pests.)

Often, people in such societies had to adjust their behavior to the natural forces – substantially outside their control – to which they were subject. The payment of respect to those forces (for example, by ascribing "spirit" to animals, plants, or environments) is an understandable extension of this situation. While behavioral change in response to feedback signals could and did occur in such societies, there was no necessary period of extensive thought between failure and correction (Alexander, 1964: 49–51). Moreover, unnecessary (and potentially destabilizing) behavioral changes were ruled out by myth, superstition, tradition, or taboo (though within certain bounds experimental adaptation to the environment could take place; see Britan and Denich, 1976). Therefore ecologically sound behavior could be the indirect result of actions with very different conscious motivation. For example, a belief in sacred cows arguably contributes to the integrity and life-support capacity of rural Indian ecosystems (Harris, 1966).

Contemporary societies have, quite simply, lost their innocence. For better or for worse, selfconsciousness about individual and collective choices is the norm. To return to the unselfconsciousness of traditional or preliterate society is simply impossible, even if we should choose to mimic the practices of those societies.[6] So, for example, Owen (1973) believes the weeding practices of African farmers may be applicable to English horticulture, but English gardeners could only choose such practices consciously.

There is a further consideration which prevents the jettison of self-consciousness and reason at this juncture. Preliterate societies may possess excellent negative feedback devices in terms of instantaneous reaction damped by tradition, but they lack the resilience necessary to cope with any severe disequilibrium conditions in the relationship between human and natural systems. One should not place too much faith in the capabilities of the indigenous peoples of the Sahel to reverse a desertification process - even if, left alone by outside influences, they would not have let that process occur. Famine in the Sahel yields a picture of helplessness in the face of disruptive environmental forces.

[6] Further, any such return would have to embody a logical contradiction, for it could be reasoned through only on a self-conscious basis.

Tradition and myth may have been functional for the survival of preliterate anarchies. Similarly, the more persistent intentional communes in contemporary industrial societies have often enforced common purpose through allegiance to a religion, or a charismatic leader. For example, the long-lived Hutterite agricultural communes in the USA and Canada are governed – at least notionally – by divine revelation, which imposes a degree of simplicity and frugality on the lives of their members (see Bullock and Baden, 1977). But myth, tradition, religion, and charisma clearly destroy the "openness" required for resilience.

If a locally self-sufficient form of social choice – indeed, any form of social choice – is to be resilient, then it must be governed by reason (either instrumental or practical) rather than tradition or myth.[7] Preliterate anarchies (or, for that matter, Rousseau's ideal rustic republic) could do without reason, but we cannot. Autonomy and self-sufficiency assist here by concentrating the problem-solving mind, for disequilibrium with the local environment means that the members of a community have some clear and urgent tasks. The community's energies cannot be put to devising ways of offloading their problems onto distant ecosystems and societies (for example, by importing large quantities of food, or exporting toxic or radioactive wastes to the Third World). Self-reliance imposes a form and content upon feedback and problem-solving (notably lacking in the standard open society).

Local autonomy and self-sufficiency do not guarantee that a community will not seek instrumental control of local ecosystems, and pursue ends at odds with ecological integrity – despite the immediacy of negative feedback signals. So, for example, the kibbutzim of Israel are not noted for their ecological sensibilities; for the most part, they engage in agriculture which is intensive in its use of energy and artificial chemicals (Allaby and Bunyard, 1980: 212–13). And this is why "small is beautiful" needs to proceed in concert with practical reason if it is to promote ecological rationality.

COOPERATION AND COORDINATION

It is indicative of the extent to which we have been conditioned by Hobbesian thinking that the word "anarchy" is associated in the popular

---

[7] Murray Bookchin seems to suggest that anarchical societies can embody resilience in exactly the same way that ecosystems do – in a spontaneous tendency to become more differentiated and complex with time (succession). Indeed, he believes that the measure of progress in both ecosystems and social systems is differentiation (Bookchin, 1982: 30–2). It may be appropriate to allocate resilience to an unselfconscious process of this sort in preliterate anarchies, but Bookchin's suggestion is clearly inadequate in societies that have lost their innocence.

imagination – indeed, in the imagination of most social scientists – with disorder or chaos. This equation notwithstanding, any radically de-centralized design can, in fact, draw upon a number of coordinating devices, none of which involves recourse to the mechanisms discussed in chapters 7–11 of this study.

Small autonomous communities – whether preliterate anarchies or (less likely) intentional communes – have often managed to maintain a high degree of order, stability, and social control without any hint of hierarchy or formal authority at the system level (see, for example, Roberts, 1979). An absence of *central* control is no proof of the absence of control *per se*. Order can be maintained in anarchical communities through devices such as a general fear of violent conflict, ostracism and ridicule of those who transgress norms (Taylor, 1982: 81), shared myths, and common socialization. Such devices can operate to restrict abuse of resources held in common, or to encourage contribution to environ-mental public goods, by dramatically reducing the payoffs associated with "defection" in the prisoner's dilemma.

The general lesson here is that decentralized controls which secure compliance are a real-world possibility (see also Young, 1979). So, for example, the technical anarchy of the international system possesses controls such as the use of force by individual actors (and a concomitant fear of that force), felt obligations on the part of members of govern-ments and other collective actors, and convergence of expectations around agreements. The precise content of any decentralized controls must, one suspects, be case-specific. Hence, for example, punishment of transgression through economic sanction (such as withholding potentially beneficial trades from an actor) might work in some cases, but less readily among actors who set little store by narrowly economic values.

Aside from the existence of social controls of this sort, it remains the case that, even for narrowly self-interested actors, cooperation with other actors can be a thoroughly rational strategy to pursue in a radically decentralized system.

This contention would appear to fly in the face of the formal logic of the prisoner's dilemma (discussed in chapter 4 above). The traditional analysis of common property problems and public good underprovision (for example, Hardin, 1968; Olson, 1965) is incomplete, though, because it is static. That is, this analysis assumes a "one shot" prisoner's dilemma game, in which each actor encounters the other(s) on a single, isolated occasion. The real world of common property and public goods is dynamic. This dynamism means that, in a succession of choices about whether to cooperate or defect, individuals have the leeway to adopt *conditional* strategies: they can make their choices contingent upon the behavior of other actors in previous iterations.

In the static case, it is always instrumentally rational to defect by maximizing one's take from the commons, or refusing to contribute to a public good (see chapter 4). In the dynamic case, though, greater returns can be expected in the long run through the adoption of a conditionally cooperative strategy – that is, making one's own good behavior contingent upon the good behavior of others. (Any individual adopting an unconditionally cooperative strategy can expect to be taken advantage of by others.) Assuming that all actors exhibit like instrumental rationality, the dynamic case can converge on an outcome of universal cooperative behavior. This finding has been demonstrated in terms of deductive logic by Taylor (1976), and corroborated by computer simulations (Axelrod, 1984) and experimental work with human subjects.

Taylor's demonstration is based on the idea that, over successive iterations of the prisoner's dilemma game, actors can experiment with different strategies (for example, unconditional non-cooperation, various forms of conditional cooperation). Axelrod (1984) has shown that the strategy with highest long-term expected value to each individual is a simple "tit for tat."[8]

One can also argue that conditionally cooperative behavior is in a sense "natural." Recent work in evolutionary biology (for example, Axelrod and Hamilton, 1981) suggests that behavioral strategies of competition and cooperation are themselves subject to natural selection, and that strategies of high survival value will in fact contain large elements of conditional cooperation (see also Alexander, 1974).[9]

Conditional strategies of the sort that make cooperation (and therefore coordination) possible in a radically decentralized system of instrumentally rational individuals are facilitated by smallness in size of the group in question. In part, this is because cooperative outcomes are promoted to the extent the interaction of any two actors is repeated (i.e., the game is truly iterative). In addition, the adoption of a conditional strategy by any actor is possible only in the context of a knowledge of the

---

[8] Taylor shows that the prisoner's dilemma does not necessarily characterize public good and common property dilemmas – it is only in the worst case that the dilemma holds (Taylor, 1976: 25). The fact that Taylor's conclusions apply even in that worst case adds to their power.

In passing, it should be noted that conditional cooperation offers no solution to public good underprovision or the tragedy of the commons in market systems, for *competitive* striving is central to the operation of markets.

[9] This result is, of course, inconsistent with a crude Darwinian perspective, which regards competition as the inevitable norm.

prior choices of other actors. Thus the likelihood of convergence on a cooperative outcome is in inverse proportion to the number of actors in a "game" (Taylor, 1976: 92–3; see also Hardin, 1982: 180–6). This conclusion is yet another reason why "small is beautiful" in radically decentralized social choice.

Cooperation and coordination are, then, possible among even instrumentally rational maximizers in an anarchy. But there is no reason why self-interested maximization must prevail under this kind of social choice. For even if at some (hypothetical) outset actors are instrumentally rational egotists, with time the adoption of conditional strategies and concomitant achievement of cooperative outcomes to the iterated prisoner's dilemma game might be expected to facilitate a *culture* of cooperation (Taylor, 1982: 51–2).

A culture of cooperation reaches its zenith in the idea of *community*. A community in the strong sense exhibits a commonality of values and beliefs, direct and many-sided relationships among individuals, and reciprocity manifested in altruism (Taylor, 1982: 26). The notion of community, and an absence of market social choice, distinguish the kind of anarchism developed here from libertarian individualism or "anarcho-capitalism" – as proposed, for example, by Nozick (1974). Market relationships entail a division of labor and rank which strengthens particular interests at the expense of common concerns.[10] It is the spirit of cooperation which moves radical decentralization of the kind sketched here away from the market and toward practical reason.

Kenneth Boulding notes that an obvious non-authoritarian solution to the tragedy of the commons is the "comedy of community" (Boulding, 1977: 286). The tragedy of the commons is rare in groups that are organized on a communal, cooperative basis. So, for example, deforestation in the tropics tends to occur only when cooperation and community break down, either from within or under the pressure of external forces such as the market economy (Shiva, Sharatchandra, and Bandyopadhyay, 1982: 161–2). When this culture does break down, even small self-reliant settlements are quite capable of destroying the environment on whose integrity they ultimately depend (see Eckholm, 1982).

Community and the cooperation and coordination it fosters are again facilitated by smallness of size in social organization. The smaller a group,

[10] The debilitating effects of a division of labor constitute one reason for Rousseau's (1968) belief that his republican Utopia could be realized only in small, pre-industrial peasant communities. As already noted, Rousseau had little faith in the ability of ordinary people to deal with complex problems.

the easier it is for individuals to identify with it as a community (Edney, 1981).

Cooperation, community, and the decentralized generation of common norms make social order and coordination possible in an anarchy. Hence anarchy does not dispense with order and organization. Order and organization of this sort in turn make communal property practicable. Under communal property, access to the productive or waste-assimilative capacities of land or ecosystems is effectively regulated and decided upon by the community as a whole.[11]

The potential of a radically decentralized design in terms of negative feedback and coordination is such that questions of flexibility and robustness now come to the fore. Robustness is, in fact, highly problematical (see the next section). Flexibility might be more readily attained: the libertarian spirit is one of improvization and experimentation. Moreover, there is little in the way of formal structure at the system level to change, or stand in the way of change. The social controls that necessarily exist in a decentralized system can be diffuse and malleable, rather than cast in the concrete of formal institutions.[12] However, the degree of flexibility attained would depend upon the specifics of the social controls in operation, about which it is hard to generalize.

Overall, then, it would appear that a radically decentralized design can be sketched which performs adequately on the criteria of negative feedback, coordination, flexibility, and resilience. This tantalizing vision faces several obstacles.

IMPRACTICAL ANARCHY?

The first major obstacle to the radically decentralized vision falls under the general heading of robustness. Any small-scale, autonomous and communal social structure can find it hard to flourish in a world of *states*.

Large-scale anarchistic experiments (for example, Paris in 1871, the Ukraine in 1917, Spain in the 1930s, and Hungary in 1956) have not shown much in the way of staying power. No large-scale intentional

---

[11] Private property can, though, be retained to a degree. Bookchin suggests that any such property could be held in usufruct (Bookchin, 1982: 50–1). Schumacher urges a distinction between "property that is an aid to useful work" and "property that is an alternative to it" (Schumacher, 1973: 263).

[12] Social controls in traditional and preliterate societies may constitute an exception here.

anarchy (with the possible exception of France in May, 1968) even had the chance to collapse from within: all were vanquished by force of arms.

Historically, the authority of states grew largely because a state can easily take over any preliterate anarchy in its vicinity or colonizing path. The only escape for the threatened society was to organize *itself* into a state (Taylor, 1982: 130); a similar process may be observed in the militarization of the Israeli kibbutzim. Less dramatically, the intervention of government in previously self-reliant communities can destroy their feedback sensitivity. McCay (1978) documents a case in which government intervention in fisheries management in Newfoundland led fishermen to undertake practices they knew full well would deplete fish stocks; the fishermen involved started to defect rather than cooperate, and hence ceased to be a self-regulating group.

A kind of Gresham's law threatens communal forms: a radically decentralized cooperative structure faces peril in a world of states. The question here, then, is less whether "small is beautiful" than whether "small is possible." How may radically decentralized innovation in social choice plausibly proceed? Schumacher's (1973) suggested strategy is the gradual population of the world with appropriate technologies (appropriate in scale and simplicity to the needs and capabilities of the people who are to use them), cooperative industrial enterprises, and small-scale, diversified, locally controlled economic organizations. Thus, Schumacher portrays in glowing terms the Scott–Bader Commonwealth in England. Scott–Bader is a worker-controlled enterprise which makes decisions on a cooperative and decentralized basis – yet still manages to flourish in a market economy. What Schumacher commends is essentially a process of structural renewal from the bottom up.

Along these lines, an ambitious experiment was undertaken in the Adams–Morgan neighborhood in Washington, DC – a community of 40,000 people – over a five-year period in the early 1970s. This experiment involved the promotion of self-sufficiency in food production, services, and utilities, together with government by participatory "town meetings" (see Hess, 1979).

On a still larger scale, programs to promote community-controlled "social forestry" are currently in vogue in a number of Third World countries. The intention of these programs is to promote reforestation to correct for fuelwood shortages and environmental damage caused by deforestation (Barnes and Allen, 1981). For the most part, social forestry is sponsored by governments and international aid agencies.

Such islands of community and cooperation do indeed exist in today's world, if they are a little thin on the ground. Any more widespread application of these principles threatens all forms of organization which embody hierarchy, whether they are states, business firms, or legal systems. Therefore substantial opposition is likely, in the form of either subversion or outright repression.

There is, too, a more subtle means through which piecemeal communitarian designs may be nipped in the bud. For the very existence of a state makes it the focus of demands; people call upon the state and its agencies to do things, instead of helping themselves or each other (see Gorz, 1980: 34). Thus formal government, by its mere existence, can lead to the atrophy of cooperation and altruism. By the very scale of the society over which it provides, and through a general promotion of "largeness" in social organization, the state makes itself more necessary (Taylor, 1976: 133–6).[13] Thus participants in any embryonic decentralized experiment face a constant temptation to turn to government for help as soon as they run into difficulties. Moreover, at least in contemporary polyarchies, government may readily oblige with monetary assistance. The Adams–Morgan experiment mentioned above foundered in part for this reason, as the individuals involved diverted their energies away from self-sufficiency and toward the attainment and administration of government grants. A similar obstacle to self-sufficiency may be discerned in the remote communities of Alaska, where the more able individuals in communities devote a great deal of time and effort to securing government finance, thereby perpetuating dependence on outside actors and forces (see Dryzek and Young, 1985).

Experiments in radical decentralization do, then, face hazards in a world of states. Some might argue that this peril may be waning, inasmuch as there are signs that in today's world the state system may itself be weakening, thus removing one obstacle to the decentralized vision (see Falk, 1983: 277–98). Whether or not this weakening trend really exists remains a controversial question. If there is no such trend, opportunities do remain for decentralizing innovation (as will be argued in the final chapter). Assuming for the moment that the trend does exist, the interesting question here would concern what replaces the state.

If the state's replacement is the market, one would expect continued subversion of communal and cooperative designs. To illustrate, in India any potential positive effects of social forestry on rural ecosystems have been thwarted largely by the commercial imperatives of the market. Reforestation proceeded guided by the commercial potential of the trees planted, rather than community or ecological effects (see Shiva, Sharatchandra, and Bandyopadhyay, 1982).

[13] As noted above, strategies of conditional cooperation become less likely as the size of a social unit grows.

The second major oversight in the decentralized vision concerns coordination above the local level. It is clear that cooperation without central authority can flourish most readily in small-scale communities. As community size increases, cooperation and social order become more problematic. This recognition might at first seem to add to the case for decentralization. Unfortunately, it is not easy for even the most self-sufficient local communities to remain hermetically sealed from one another. Moreover, many contemporary ecological problems do, of course, transcend the local level. Problems such as acid rain, desertification, oceanic pollution, radiation, and the "greenhouse effect" are regional or even global in extent. The question of coordination above the local level therefore looms large, and this is the point where decentralizing theory is at its weakest. The standard anarchist answer is to postulate loose "confederations" of communes, and, at still higher levels, "confederations of confederations" (see, for example, Ward, 1977).

A more specific answer in the context of organization theory is provided by Thayer (1981). Thayer proposes that governmental hierarchies be supplanted by series of "overlapping circles" from local to national levels. Each "circle" is a small group operating on a consensual basis. Neighborhood groups would contribute members to city groups, which would in turn contribute members to national groups. Hence consensus at the national level would reflect a kind of "collective will," congruent with consensus at all lower levels (see Thayer, 1981: 171–4). Unfortunately, there is no guarantee that anything more than stalemate would result from Thayer's scheme. One suspects that any such structure would lapse into either hierarchy or bargaining.

Under confederation – or, for that matter, Thayer's overlapping circles – the number of actors participating at higher levels of choice is still fairly small (as indeed it is in the contemporary international system – see chapter 12). The coordination problem lies not so much in the number of actors as in the fact that each actor is itself a collective entity. Therefore many of the social controls and other mechanisms that can effectively secure cooperation and coordination within small communities can have little effect when it comes to relationships between communities (see Taylor, 1982: 167). Collective entities cannot feel moral obligation, or suffer ostracism, or share a sense of community, or even adopt consistent conditional strategies.

CONCLUSION

Radical decentralization has substantial promise in terms of its mapping of ecological feedback signals onto social choice, its facilitation of practical reason in collective deliberations and actions, and its coordination potential at the communty level. Its most problematic feature is its fragility in a world of states. Radical decentralization is largely silent on coordination in ecological problem-solving above the local level. Mechanisms for the achievement of this important kind of coordination which do not sacrifice practical reason or decentralization form the topic of the next chapter. Despite its flaws and gaps, radical decentralization – especially when combined with practical reason – is suggestive of improved ecological rationality in social choice.

17

Two Tactics for Coordination

Ecological problems do not respect political boundaries. A city, commune, region, or country may be adjudged an ecological failure if it does not allow and promote coordination across boundaries in response to joint problems. This chapter will explore means for such coordination.

The difficulty of achieving coordination across boundaries is, if anything, likely to be magnified by any institutionalization of the innovations sketched in the preceding chapter. The substantial autonomy concomitant with radical decentralization would seem to imply a minimum of supra-local social controls.

There does seem to be a paradox here. Coordination among social actors – whether through conditional cooperation, decentralized social controls, or discursive consensus on norms – is clearly facilitated by smallness of size in the social unit. This small scale leads, *ceteris paribus,* to the existence of large numbers of such units. And the larger the number of social units, the more problematical is coordination above the local level. For strategies of conditional cooperation become less rational as the number of actors increases; defectors are less conspicuous; and decentralized social controls lose much of their force.

The challenge at this juncture is to explore the possibilities for effective large-scale coordinating devices which will simultaneously preserve the decentralization necessary for effective negative feedback and for resilience. One can rule out administrative and legal mechanisms at the outset, for they reek of centralization. But markets and polyarchy may be worth another look.

MARKETS AND POLYARCHY

The market achieves coordination among actors through signals transmitted via the medium of the price system. Clearly, the price system can operate effectively even when the number of actors involved is very large, or when these actors are themselves collective entities.

Could one supplement social choice structures embodying substantial local autonomy and practical reason with a market system for supralocal coordination? Such a vision of internally participatory forms connected through a market has its attractions to those who espouse "market socialism" (see Lindblom, 1977: 330–43; Goodin, 1982).

Markets fail with respect to the coordination desideratum for reasons associated with the prisoner's dilemma (see chapter 7). While this problem is the bane of any decentralized form of social choice, there are, as noted in chapters 15 and 16, a number of ways in which it can be surmounted. The pertinent devices here are conditional cooperation, commitment to community, decentralized social controls, and practical reasoning. Community, social controls, and practical reason obviously have no place in a market order; but what of conditional cooperation? Clearly, cooperation of sorts does occur in the systems we ordinarily classify as markets. Examples would include cartels, tacit processes of price leadership, market-sharing understandings, and so forth. However, such arrangements really do step outside the domain of market social choice, bearing a stronger resemblance to bargaining or administration. Moreover, they are feasible only if the number of actors is small.

One must conclude that the market *qua* market possesses absolutely nothing in the way of coordinating devices to cope with the prisoner's dilemma, and hence with common property or public goods problems. The best the market can offer is the offloading of these problems onto other forms of social choice. Indeed, given the tendency of markets to "imprison" the forms of collective choice with which they co-exist (see chapter 7), markets are a positive obstruction to coordination through other means. Nor can one adopt the coordinating aspect of the market in isolation, for accepting the price system is equivalent to accepting the market.

Polyarchy fares slightly better than the market. Coordination across *decisions* in a polyarchy is achieved through decentralized mutual adjustment and reciprocal adaptation. Coordination across *actors* can involve mutual adjustment too, although explicit bargaining is also important. Clearly, these devices allow considerable scope for conditional cooperation and hence coordination.

Two factors prevent an endorsement of polyarchy's contribution to coordination here. The first obstacle is that polyarchy's coordination processes cannot easily be plucked wholesale and grafted onto the innovative forms of social choice sketched in the previous two chapters. For polyarchy's coordination devices require, *inter alia*, some clear locus of governmental authority at the system level, a set of rules to govern interactions and identify legal authority, and a variety of checks and balances requiring in their turn a degree of complexity in the interactions of collective actors. Polyarchy, like the market, can be adopted only in its entirety. It is hard to imagine a radically decentralized system embodying practical reason coordinated by polyarchy.

The second obstacle, still more serious, is that polyarchy's claims to coordination in an ecological context simply do not stand up. Under polyarchy, one can expect an arbitrary decomposition of essentially non-reducible ecological problems. (One example stressed in chapter 9 was the energy problem in the USA.) The consequence is fragmented and incoherent response, with little likelihood of convergence on a coordinated outcome. Polyarchy too has little to offer in terms of supra-local coordination.

FIRST TACTIC: LIMITED BARGAINING

The summary verdict on bargaining in chapter 12 was negative. Why, then, resurrect bargaining at this juncture?

Bargaining is cumbersome. Feedback channels are likely to be extremely clogged if all feedback signals must pass through a negotiating forum before any response can be made. Lightning reaction to changed conditions is unlikely. Bargaining was sunk by its poor performance on negative feedback and resilience. Its potential for coordination is roughly equivalent to that of polyarchy. But, in contrast to polyarchy, bargaining's coordinating device – formal negotiation – can be grafted onto other forms of social choice without necessarily displacing them.

Under bargaining, there can be a sizable "contract zone" of coordinated outcomes, especially when the alternative is a chaotic Hobbesian free-for-all (see chapter 12). Determinate outcomes are most easily achieved if the number of actors remains small. As the number increases, so do the "transactions costs" of reaching agreement – though one should note the success of the United Nations Conference on the Law of the Sea in (tortuously) negotiating the assent of nearly all the world's sovereign states. But the likelihood of subsequent compliance with accords also declines with the number of actors involved, as free riders have more incentive to break loose.

The capacity of a bargaining system can soon become stressed as the number of issues to be dealt with rises. This consideration might reinforce the case for radical decentralization discussed in chapter 16; for, the greater the degree of local autonomy, the fewer the issues demanding attention above the local level (other things being equal). The paradox here is that, the greater the degree of local autonomy, the larger the number of actors to be expected at a bargaining table, and the more difficult it becomes to secure agreement and coordination. Whether or not a workable compromise exists between the twin requirements of "few actors" and "little to decide about" is a matter for empirical test.

Nonetheless, the coordinating potential of bargaining is greater than that of markets and polyarchy. Unlike the market, bargaining does not export its imperatives to other forms of social choice. And unlike both markets and polyarchy (see above), it can be adopted partially.

SECOND TACTIC: PRACTICAL REASON BY FUNCTIONAL AREA

In chapter 15 it was noted that communicative rationalization has considerable coordinating potential in small-scale groups. Is there any possibility that supra-local coordination could be achieved in like manner? Following Jean-Jacques Rousseau, many enthusiasts of discursive participation would reply in the negative. To Rousseau, direct democracy could thrive only in small communities facing simple problems. Complex problems – especially trans-local ones – would be the death of direct democracy. In this sense, Rousseau anticipated Max Weber, who portrayed bureaucratization as an inevitable response to rising social complexity (see especially Weber, 1968).

More positively, one possibility here might be general-purpose institutions (analogous to the legislature of a national political system) governed by practical reason. But more realistic than any such broad-ranging institution – and more in keeping with maximum local autonomy – would be communicatively rationalized mechanisms for dealing with problems in specific functional areas (such as acid rain, fisheries management, nuclear wastes, and so forth).

Participants in any forum of this sort would be those with a stake or interest in the problem area in question. Whether participants would be representatives from locally constituted groups or unattached individuals is an open question.

Ideally, a communicatively rationalized process of this kind could achieve supra-local coordination through the development of intersubjective understanding and perhaps even commitment to shared norms upon which actions (individually or in concert) would be based. This is not mere wishful idealism. It is noteworthy that, at the height of their countries' conflict over the Falkland Islands, representatives from Britain and Argentina were taking part in negotiations in New Zealand over the future of Antarctica. Similarly, Israel and the relevant Arab states have participated together on the Mediterranean Action Plan (sponsored by the United Nations Environment Program), designed to cope with pollution in the Mediterranean basin. In each case, a commitment to shared (generalizable) environmental values and an awareness of the need to reach reciprocal understanding overcame hostility on other issues.

The constitution of groups governed by practical reason and defined by problem area is not completely novel. Organization theorists and business managers have occasionally recognized the problem-solving capacities of small groups operating in rough approximation to principles of free discourse. So business firms sometimes find it expedient to hand especially tough questions – particularly those demanding creative solutions to novel problems – to collegial groups outside the organizational hierarchy (Bruner, 1962).

The difference between these organizational cases and the kind of innovations under study here is, of course, that business firms and other large organizations generally set narrow limits upon goals and values. Discursive and participatory forms are merely instrumental to the larger purposes of the organization. Hence the success of collegial forms in an organizational setting does not guarantee like success in cases where values are less clear-cut.

Nevertheless, communicative rationalization has been pursued with some success in cases characterized by conflicting interests and values. The environmental mediation attempts discussed in chapter 15 clearly fall into this category. Those cases were generally local and site-specific; however, several attempts have also been made to apply the spirit of environmental mediation to large-scale, functional-area problems. I will now discuss two such cases, from the USA and Canada, respectively.

American Coal Policy

The National Coal Policy Project, which operated in the USA from 1976 to 1979, constitutes perhaps the first application of the general principles of mediation to a broad functional-area environmental problem. The Project brought together representatives from industry (both coal-producing and coal-consuming) and national environmental groups, under the neutral auspices of Georgetown University's Center for Strategic and International Studies.

A number of plenary sessions were held over the life of the Project, but the bulk of the work was done in committees dealing with air pollution, transportation, mining, conservation, and so forth. Plenary sessions and committees were all bound by a set of guidelines adopted wholesale from Wessel's (1976) "rule of reason." That "rule" is a set of principles for reasoned debate, incorporating, for example, prohibitions on concealment, delay, dogmatism, withholding of information, *ad hominem* attacks, and the like.

The participants in the Coal Policy Project did, apparently, abide by the rule. The product was a substantial measure of agreement on the future of coal mining, transportation, and use, reflected in a detailed and lengthy final report (Murray, 1978).

The mere fact of agreement is not, though, indicative of any exercise of practical reason, or of any promise of coordination in the sense required by ecological rationality. Some of the agreements reached in the National Coal Policy Project were clearly little more than fairly expedient tradeoffs between competing particular interests. So, for example, the environmentalist participants consented to a simplified "one-stop" permitting procedure for coal-burning plants, in return for industry's support for public financing of environmental groups in the relevant hearings.

Coordination through discursive social choice would ideally reflect a measure of consensus on generalizable values. No such clearly *ecological* values came to the fore in the Coal Policy Project discussions. However, one striking feature of the Project was the extent to which all the participants adopted the language of microeconomics, embracing concepts of market principles, externalities, public goods, and the like (see McFarland, 1984). A large number of the recommendations were couched in similar terms – for example, a suggestion that existing regulatory mechanisms for pollution control be supplanted by a system of emission standards and charges. The "generalizable" interest that came to the fore was the welfare economist's favorite: social welfare in allocative efficiency terms. While that standard leaves much to be desired in ecological terms (see chapter 5), its claim to generality is clearly much stronger than the interests typically found in polyarchical interaction. Microeconomic efficiency principles generally fare very badly in the ordinary processes of polyarchy (see Wildavsky, 1966). Therefore, although "efficiency" is not very attractive ecologically, the Project's recommendations clearly did constitute a coordinated set through reference to that value. The kind of coordination obtained constitutes a substantial improvement over the piecemeal products of polyarchy.

Beyond notional agreement, effective coordination requires that any accords be acted upon. Unfortunately, the National Coal Policy Project's recommendations have not been reflected in public policy. The first reason for this failure stems from the exclusion of a number of interested actors: the United Mine Workers, consumer groups, and – crucially – two environmental groups who subsequently denounced the accords. Those environmental groups who had participated and consented were forced to retreat for fear of endangering environmentalist solidarity. The second reason for failure was that no effort was made to involve governmental actors (such as federal agencies) in the Project. Such actors have strongly held interests of their own, which were not accommodated (see McFarland, 1984, for greater detail).

Two morals can be drawn from the implementation failure of the Coal Policy Project. The first is that the effectiveness of any discursive structure depends on the inclusion of all interested actors. The second is that co-existence with other forms of social choice can be problematic. The imperatives of the polyarchical (and legal) "game" affected the subsequent stance of the environmentalists, and the existence of an administrative regulatory structure impeded acceptance of the Project's recommendations.

One suspects, though, that the two implementation problems encountered by the Project are eminently solvable, given a little foresight. Hence, despite its ultimate failure, the experience of the National Coal Policy Project is an encouraging one.

Canadian Pipelines

One of the failings of the National Coal Policy Project was restricted participation. No such restriction was made in my second example, which in this respect approached the discursive ideal more closely. The MacKenzie Valley Pipeline Inquiry, conducted under the auspices of Justice Thomas Berger between 1974 and 1977, constitutes a discursive form which can be located midway on a spectrum between "site-specific" and "functional-area." The mandate of the Inquiry was, initially, narrow: to investigate the potential social, environmental, and economic effects of proposed pipelines to transport oil and gas from the Canadian and Alaskan Arctic to southern markets. The Canadian North is a region characterized by fragile ecosystems, wilderness, substantial poverty, a weak economic base, and indigenous peoples (Inuit and Dene) torn between traditional ways of life and industrial society.

Most judicial commissions of inquiry operate in a highly formal legalistic manner or, alternatively, simply aggregate the submissions of expert witnesses. Berger, in contrast, set in motion a process of inter-action and dialogue among all the actors involved – even those with no prior political organization or experience. These actors included native people's organizations, environmental groups, various governments and agencies, the oil and construction industries, consumers, and the ordinary residents of the Canadian North. In addition, the Inquiry drew upon "experts" such as engineers, economists, ecologists, and anthropologists, both for submissions to the Inquiry and as staff members. Large numbers of hearings were held, including "community hearings" throughout the North. Finance and information were disseminated to individuals and financially weak interests to assist in the preparation and articulation of arguments.

Communication among these actors was multi-directional; the Inquiry provided a vehicle for prolonged and public discourse. This discourse covered moral questions and broad issues of development strategies for the North, as well as the more narrow factual considerations of pipeline impact. The Inquiry started with no pre-ordained normative position (other than Berger's personal predispositions), but instead appeared to develop that position as the debate proceeded.

The Berger Inquiry may represent one of the best real-world examples of communicative rationalization in public policy yet attained (see Dryzek, 1982: 324–5). Its performance on the coordination desideratum is also encouraging. Berger transcended the impact evaluation mandate to consider the normative basis and empirical content of political, cultural, and economic development strategies for the Canadian North. The formal product of the Inquiry was a coherent strategy involving a moratorium on pipeline construction pending a native lands claim settlement and the establishment of a sound renewable resource-based economy in the region (see Berger, 1977). These conclusions were accepted (if with some reluctance) by the Canadian federal government.

Berger's report met some opposition, both from those skeptical of the possibility of co-existence between "parallel economic sectors" (renewable and non-renewable), and from those whose economic interests were slighted. Yet there is no denying Berger's accomplishment in producing a coherent development strategy which met with a large measure of agreement from a very diverse range of actors. Unfortunately, market, polyarchy, and administration have since intervened to prevent the implementation of the more positive aspects of Berger's program: those pertaining to renewable resource development.

CONCLUSION

The Coal Policy and Canadian pipelines cases indicate the potential for coordination through comunicative rationalization in supra-local ecological problem-solving. Procedures of this sort can also stray (and hence coordinate) across functional areas, as both cases show. Moreover, the potential of such procedures may actually be increasing with time. In particular, the technology of the information revolution – interactive television, computer networks, telecommunications, and the like – holds out strong possibilities for communicative rationalization transcending the local community. (See Barber, 1984: 273–81 and Luke and White, 1985, for the relevant technical details.) Such developments would further contradict Rousseau's denial of the possibility of direct participation in collective choices outside small primitive communities. Conversely, they would corroborate the contention of Habermas (1984) that modernity brings increasing potential for communicative rationalization of public life, as well as a more dangerous potential for bureaucratization.

On the other side of the coin, the coal policy and Canadian cases illustrate the difficulty that practical reason can expect in co-existing with any other form of social choice, whether market, polyarchy, administration, or law. Nevertheless, bargaining and practical reason – in isolation or in juxtaposition – could contribute to effective supra-local coordination.

18

From Design to Meta-design

The agenda for institutional reconstruction emerging from the previous three chapters is clear enough. Its most prominent items are decentralization in the form of substantial local autonomy and self-sufficiency, open and discursive "communicatively rationalized" social choice, and (perhaps) limited bargaining. The designs of chapters 15–17 have, though, been depicted in very broad strokes. The details remain to be filled in. In addition, one should allow for the possibility that the innovations discussed have flaws I have not thought of. If those flaws are minor, they need to be ironed out. If they are major, then competing designs need to be articulated. The elaboration of details, elimination of flaws, and stimulation of competing designs would benefit from empirical testing. How may such testing proceed, other than through piecemeal innovations of the kind used to illustrate chapters 14–17? Can any more systematic pattern of social learning be envisaged? In other words, is there any hope for restructuring social choice mechanisms on a general level?

THE POSSIBILITY OF DESIGN

A glance at history shows that both evolutionary and revolutionary changes in the structure of social choice can and do occur. But what motivates those changes? The question of motivation is crucial, for a world beset by ecological problems would require innovation in social choice that is neither random nor law-governed.

The possibility that the development of social choice mechanisms is governed by laws of history may be dispensed with safely. Historicism has suffered enough at the hands of Popper (1961) to render superfluous anything I might add.[1] Ecological theories of history (for example,

---

[1] Marxism is the best developed and most widely known form of historicist analysis. For Marxists, of course, the movement of history culminates in communist society.

Colinvaux, 1982) are no more convincing. Such theories are based on
the biological characteristics (such as an ability to learn a niche) of
individual members of the species *Homo sapiens,* effectively ruling out
the possibility that factors such as culture and the structure of social
choice have any real importance.

An element of randomness is less easily dismissed. Randomness of
sufficient magnitude might vitiate any notion of conscious innovation in
social choice. Perhaps the best way of demonstrating the possibility that
there is more to the construction and evolution of social institutions than
mere chance is to point to some concrete historical examples of conscious
design in social choice.

There is, in fact, a large number of examples available here. The
instance which should spring to the mind of most Americans is that of the
proto-polyarchy (or Whig republic) designed by the authors of the US
Constitution. Perhaps more significant on a global scale is the establishment during the nineteenth century of the conditions for market social
choice. As Polanyi (1944) points out, this "great transformation" was a
product of conscious intervention on the part of governments. The
twentieth century has seen successful attempts to magnify the centrally
administered aspects of social choice, not only in the Soviet bloc, but also
in the wartime economies of the Western democracies. Indicative planning of the market has made significant inroads in Western Europe, and
moral persuasion has been instituted under the auspices of Mao and
Castro. One could multiply examples.

Each innovation has its own story. Some are revolutionary, some are
iterative; some are informed by theoretical works, some proceed through
atheoretical trial and error; some have predictable consequences, some
produce real surprises; some are painful, a few proceed smoothly. This
variation should come as no surprise, for each instance proceeded against
the background of a particular historical context. Pertinent dimensions
of "context" here include the existing mix of social choice structures,
opportunities for collective actions, the perceptions and values of the
individuals involved, and the constraints upon actions from sources as
diverse as threats from foreign powers, the existing constitutional order,
and the power of countervailing groups in society.

Historical analogies range only from the imperfect to the misleading;
therefore one should hesitate before drawing too strong an inference
from these examples. Nevertheless, the general moral is clear: self-
conscious innovation in social choice is a real-world possibility. Such
innovation can take two forms. We can contemplate the construction *de
novo* of rational mechanisms. Alternatively, we could start with the
present – defined in terms of the available range of social choice devices –

and seek iterative ways of enhancing ecological rationality in social choice. I will now take a detailed look at these two sorts of design – *de novo* and iterative – in the interests of deciding how much each has to contribute to the construction of ecologically rational forms of social choice. It should be noted that the innovations sketched in chapters 15–17 could be put to use in *both* of these kinds of design.

DESIGN *DE NOVO*

The more dramatic approach to the design of social institutions would start from a blank slate and seek to establish desirable structures based on primarily theoretical considerations. Proponents of the radical view might note the hazards that established forms of social choice pose to any more limited innovation. Thus, the success of environmental mediation (detailed in chapter 15) has been constrained by the dominant social choice mechanisms in the US political economy. Any corporate actors participating in mediation have profits to pursue in the marketplace, and their attitudes remain conditioned by this imperative. Environmental groups have points to score in polyarchical interaction, and possible legal actions to back. Members of local governments can be reluctant to allow a mediation forum to trespass upon their administrative prerogatives. The implementation failure of the National Coal Policy Project in the face of the realities of polyarchy (see chapter 17) is also sobering here. If coexistence is hazardous, why not simply dispense with existing forms, and start *de novo*?

At its most self-confident, *de novo* design embodies what Boguslaw (1965) calls a "formalist" methodology. Formalism entails full articulation of formal models of a system or structure, to be used as blueprints for construction. Clearly necessary here is considerable knowledge of the elements and interactions in the system under study, together with substantial predictability in their operation. These requirements are readily met in much engineering and architectural design. Even in that domain, though, the systems in question are not always highly predictable. Note, for example, the use of prototypes by less than fully self-confident engineers, some of the social failings of architectural design (Wolfe, 1981), and the treatment of computers by their engineers as empirical systems (to be discovered) rather than normative systems (to be created) (see H. A. Simon, 1981: 24–6).

The villains of the piece bedevilling formalist engineering and architectural design here are two old friends; complexity and uncertainty. Those circumstances apply, of course, *a fortiori* to designs involving human social systems, ecosystems, and especially their interactions. At the very least, formalist design in the context of social and ecological systems demands considerable immodesty on the part of the designer.

Intelligent *de novo* design in social choice would commence with a characterization of the ecological circumstances of social choice, of the kind made in chapter 3 above. Those circumstances are complexity, non-reducibility, variability, uncertainty, spontaneity, and collectiveness. These six aspects constitute the context to which a social choice design must be "fitted." All that is needed is the addition of a criterion variable – in this case, human life support – and theoretical knowledge about the way this variable is affected by the content of the interactions between social choice mechanisms and ecosystems. One would then possess the necessary ingredients for successful *de novo* design.

The problem here is that the aforementioned six circumstances characterize not only the context of the system being designed, but also the context of any would-be *designer*. That is, the designer himself is faced directly with complexity, non-reducibility, variability, and uncertainty in the very act of design.

At this point, the formalist designer may be tempted by the standard response to complex design questions: decomposition of the overall problem into sets and subsets, according to the principles described in chapter 8. So, for example, the designer might choose to decompose the overall problem into "sets" of complexity, non-reducibility, variability, uncertainty, and spontaneity. Devices to cope with each one of these sets could be specified, and those devices could then be aggregated into a social choice structure. Alternatively, one could work with the five desiderata for ecological rationality in social choice – negative feedback, coordination, robustness, adaptiveness, and resilience – and then devise structures for meeting each, and synthesize those structures into a complete mechanism.

Unfortunately, any sets defined and devices coming out of them can be expected to intersect in complex and non-reducible ways. At best, an artificial "tree" structure is imposed on the design problem (cf. Alexander, 1965). Quite simply, interactions between any conceivable subsets would be too rich.

A good example of misconceived decomposition in formalist design in social choice in an ecological context is afforded by the regime constructed under the 1976 US Fisheries Conservation and Management Act. One of the major weaknesses of that regime is that it compartmentalizes the management of marine fisheries, totally ignoring interdependencies among ocean resources. Thus, the "optimum yield" of each resource – the expressed intent of the Act – is not readily achieved. This problem reflects the logical impossibility of simultaneously maximizing more than one variable in an interdependent system.

The formalist designer, cognizant of the hazards of decomposition in a situation characterized by complexity and non-reducibility, might respond by adopting a holistic approach to design. Many of the conjectures of futures researchers fall into this category. Unfortunately, most of those conjectures suffer from the over-simplistic view of the world which haunts any holistic analysis of a complex system (Brewer and deLeon, 1983: 47). The number of potential simplifications increases with system complexity, and so therefore does the number of potential interpretations of the system. The variety of "futures research" blueprints in existence (see Cole, 1978, for a survey) attests to this conclusion. That very variety detracts from the power of each blueprint in isolation.

Any extreme variety in formalist blueprints can produce one of two consequences. The lesser evil is some kind of compromise between competing proposals. So, for example, Young (1980) notes that a "contractarian" approach to the formation of social choice mechanisms will often yield rigid forms which, should they prove inappropriate in practice, are not readily modified.

More serious is a situation in which partisans of a particular blueprint possess sufficient power to enact their vision. Any such political actor(s) will rapidly discover that human beings lack the capacity to engage in effective wholesale social engineering. In the face of failure and massive unintended consequences, the temptation would be to save the blueprint and seek scapegoats for its failure – "reactionaries," "counter-revolutionaries," and the like. Those who question the very desirability of the blueprint – and who, if they are sufficiently numerous, threaten to block its program – tempt suppression.[2] Outside of a revolutionary situation, though, possession of the power to enact wholesale change is rare.

In attempting to salvage the possibility of a controlled transition, the formalist designer might contend that any lack of integrated design capacities should be rectified. This kind of logic leads ineluctably to centralized, authoritarian forms of government. The thrust of wholesale advocacies of blueprints is inescapably authoritarian and repressive (Popper, 1966). Those with an authoritarian bent to begin with (e.g., Ophuls, 1977; Hardin, 1977; Heilbroner, 1980) may be perfectly happy with this implication; those with leanings toward other forms of social

---

[2] As Cotgrove (1982: 103) points out in this context, "one person's heaven is another's hell."

choice (e.g., Daly, 1977; Harman, 1976) will find this logic bad news indeed. Authoritarian control is not a form of social choice particularly conducive to the promotion of ecological values or the resolution of ecological problems (see chapter 8).

A formalist approach to design should not, then, be taken literally as a program for the construction of ecologically rational social choice mechanisms. However, such blueprints do have substantial heuristic and critical utility in highlighting the defects of existing mechanisms and indicating the broad directions which more iterative innovation might take.

ITERATIVE DESIGN

Design in the realm of social institutions and practices has no universal algorithm. The content of the design task depends upon the circumstances with which the designer is faced (Dryzek, 1983d). The circumstances of design in the present context may be characterized as follows.

(1) There is an explicit set of criteria for good designs (those developed in chapter 4).

(2) There is considerable complexity in the way in which aspects of social choice mechanisms interact with natural systems.

(3) There is limited human theoretical understanding of potential or hypothetical social choice mechanisms, especially as far as their performance on the five criteria is concerned.

(4) Our understanding of established mechanisms – especially those evaluated in part II above – is much greater than that of any hypothetical mechanisms (such as those discussed in chapters 14–17), simply because we can draw more heavily on the lessons of experience.

(5) As designers, we do not have the luxury of a period of grace in which to construct any novel mechanism. As Tribe (1974: 1340–1) notes, we are like Schiller's mechanics, who cannot allow the wheels to run down as they repair "the living clockwork of the state," or like Neurath's sailors, who must rebuild their ship while at sea.

These conditions are clearly sufficient to cast doubt upon the efficacy of formalist approaches to design. The alternative to formalism dictated by complexity, limited theoretical understanding, and the need to proceed in terms of "running repairs" can only be some form of iterative trial and error, with thorough empirical tests of plausible suggestions for institutional restructuring. Experience as well as reason can be our guide. The difficulty of the conscious design task is substantially lower in the vicinity of existing mechanisms and institutional experiments than in the area of formalist projections.

META-DESIGN

At this juncture, it may safely be stated that the redesign of social choice mechanisms is indeed a real-world possibility, and that some iterative process of trial and error is necessary in any such design and innovation. Experimentation with forms of social choice promising enhanced ecological rationality is therefore both possible and desirable.

However, any piecemeal introduction of innovative forms of social choice into a world of ecologically irrational mechanisms is perilous. For example, markets "imprison" governmental social choice; a legal system formalizes social interactions beyond its bounds; and the existence of administrative prerogatives and polyarchical imperatives can undermine discursive innovations. Given these difficulties, how may any process of institutional innovation get off the ground? In other words, what are the necessary preconditions for institutional redesign?

It should be obvious here that these preconditions themselves reside in large measure in the conditions of social choice. That is, the establishment and maintenance of a capacity for institutional innovation and design is itself a pressing problem in the design of social choice mechanisms. The focus therefore now shifts from design to "meta-design" in social choice: that is, to a structural capacity to engage in innovation and design. It follows that one central consideration in the contemplation of any institutional innovation should be the capacity of that form of social choice to hold itself up to examination and, if necessary, to engage in structural modification – or perhaps even in its own supercession.

That capacity is clearly lacking in many of the social choice mechanisms discussed in part II of this study. For example, administered systems have a remarkable capacity to frustrate structural change. The inhabitants of administrative structures rarely acquiesce in change or succession of institutional forms; indeed, governmental organizations have a longevity verging on immortality. Administered systems therefore compound their ecological irrationality by securing their own perpetuation.

Law, too, is largely incompetent when it comes to institutional redesign, sharing the rigidity of administration. Legal systems generally engage in the articulation and elaboration of a legal order, not in its supercession.

In contrast to the rigidities inherent in administrative and legal structure, markets are fluid devices which can evolve – and dissolve – with ease. However, succession in market social choice follows an economically rational logic; no self-conscious institutional "search" can take place. Moreover, any modification that does occur refers only to the mix of goods and services produced. Markets are, in fact, highly resistant to change in their essential structure. Worse still, markets constrict any process of institutional redesign in the governmental choice mechanisms with which they co-exist; as noted in chapter 7, markets automatically punish any governmental actions that threaten profitability or market confidence. Note, for example, that one of the factors inhibiting practical discourse in the National Coal Policy Project discussed in chapter 17 was that the industry participants were all conditioned by the imperatives of profitability in the (imperfect) marketplace.

The central project in meta-design is the identification of forms of social choice which will facilitate their own supercession – not through collapse (as might characterize moral persuasion) or through degeneration into bureaucratic hierarchy (a fate that often befalls revolutions), but through the exercise of self-conscious collective choice. In contrast to the major existing forms, the kinds of social choice discussed in part III might promote this kind of capacity.

First of all, ideas for structural change are readily generated – if less readily acted upon – in the open society. (This study was itself conceived of and written in two of the world's most "open" societies.)

Second, the spontaneity and libertarian spirit of radical decentralization *might* mean that largely autonomous communities could be free to experiment with evolving institutional structure.

Most promising is meta-design under practical reason. Institutional forms can themselves be subjected to discursive evaluation. In communicatively rationalized social institutions meta-design is always a possibility – indeed, is always likely.[3] Practical reason both allows and requires individuals to reconstruct their relationships with one another.

The practical problem remains, though, one of setting this discursive, experimenting spirit in motion. Pending ecological crisis that might spur more decisive movement, the best that one can hope for may be the piecemeal introduction of decentralized and discursive social choice structures. Innovations of this sort can make some immediate contrib-

---

[3] As Habermas puts it in his own obscure way, "Which types of organization and which mechanisms are better suited to produce procedurally legitimate decisions depends on the concrete social situation . . . I can imagine the attempt to order a society democratically only as a self-controlled learning process" (McCarthy, 1978: 331)

utions to ecological rationality in social choice; but their real contribution may be in terms of their promotion of the *preconditions* for more
substantial institutional innovation directed toward enhanced ecological
rationality. For in remaking our institutions, we also remake ourselves:
who we are, what we value, how we interact, and what we can
accomplish. And it is for this reason above all that such experiments
merit an endorsement.

Bibliography

Ackerman, Bruce A. and William T. Hassler (1981), *Clean Coal, Dirty Air*. New Haven, Conn.: Yale University Press.

Alexander, Christopher (1964), *Notes on the Synthesis of Form*. Cambridge, Mass.: Harvard University Press.

Alexander, Christopher (1965), "A City is Not a Tree." *Architectural Forum* 122 (1 and 2): 58–61 and 58–62.

Alexander, R.D. (1974), "The Evolution of Social Behavior." *Annual Review of Ecology and Systematics* 5: 325–83.

Allaby, Michael and Peter Bunyard (1980), *The Politics of Self-Sufficiency*. Oxford: Oxford University Press.

Allison, Graham T. (1971), *Essence of Decision: Explaining the Cuban Missile Crisis*. Boston: Little Brown.

Almond, Gabriel A. and Sydney Verba (1963), *The Civic Culture*. Princeton: Princeton University Press.

Amy, Douglas J. (1983), "The Politics of Environmental Mediation." *Ecology Law Quarterly* 11: 1–19.

Anderson, Frederick R., et al. (1977), *Environmental Improvement Through Economic Incentives*. Baltimore: Johns Hopkins University Press for Resources for the Future.

Anderson, James E. (1979), *Public Policy-Making* (2nd edn). New York: Holt, Rinehart, and Winston.

Anderson, Terry L. and P.J. Hill (1975), "The Evolution of Property Rights: A Study of the American West." *Journal of Law and Economics* 12: 163–79.

Arendt, Hannah (1958), *The Human Condition*. Chicago: University of Chicago Press.

Arrow, Kenneth J. (1963), *Social Choice and Individual Values* (2nd edn). New Haven, Conn.: Yale University Press.

Ashby, W. Ross (1960), *Design for a Brain* (2nd edn). New York: John Wiley.

Axelrod, Robert (1984), *The Evolution of Cooperation*. New York: Basic Books.

Axelrod, Robert and W.D. Hamilton (1981), "The Evolution of Cooperation." *Science* 211: 1390–6.

Barber, Benjamin (1984), *Strong Democracy: Participatory Politics for a New Age*. Berkeley: University of California Press.

Bardach, Eugene and Lucian Pugliaresi (1977), "The Environmental Impact Statement versus the Real World." *The Public Interest* 49: 22–38.

Barnes, Douglas F. and Julia C. Allen (1981), "Deforestation and Social Forestry in Developing Countries." *Resources* 66: 7.

Barnett, Harold J. and Chandler Morse (1963), *Scarcity and Growth: The Economics of Natural Resource Availability.* Baltimore: Johns Hopkins University Press for Resources for the Future.

Barry, Brian (1973), *The Liberal Theory of Justice.* Oxford: Clarendon Press.

Bartlett, Robert V. (1986), "Ecological Rationality: Reason and Environmental Policy." *Environmental Ethics* 8: 221–39.

Bator, Francis M. (1957), "The Simple Analytics of Welfare Maximization." *American Economic Review* 47: 22–59.

Bauer, Raymond M. (1969), *Second Order Consequences.* Cambridge, Mass.: MIT Press.

Baumol, William J. and Wallace E. Oates (1975), *The Theory of Environmental Policy.* Englewood Cliffs, NJ: Prentice-Hall.

Beer, Charles (1982), "Environmental Education," pp. 101–15 in Dennis L. Little, Robert E. Dils, and John Gray (eds), *Renewable Natural Resources: A Management Handbook for the 1980s.* Boulder, Col.: Westview Press.

Bell, Daniel (1973), *The Coming of Post-Industrial Society.* New York: Basic Books.

Bell, Daniel (1974), "The Public Household – On 'Fiscal Sociology' and the Liberal Society." *The Public Interest* 37: 29–68.

Bell, Daniel (1976), *The Cultural Contradictions of Capitalism.* New York: Basic Books.

Berger, Thomas (1977), *Northern Frontier, Northern Homeland: Report of the MacKenzie Valley Pipeline Inquiry.* Toronto: James Lorimer.

Bernstein, Brock B. (1981), "Ecology and Economics: Complex Systems in Changing Environments." *Annual Review of Ecology and Systematics* 12: 309–30.

Bernstein, Richard J. (1976), *The Restructuring of Social and Political Theory.* Philadelphia: University of Pennsylvania Press.

Bertalanffy, Ludwig von (1968), *General Systems Theory.* New York: George Braziller.

Boguslaw, Robert (1965), *The New Utopians: A Study of System Design and Social Change.* Englewood Cliffs, NJ: Prentice-Hall.

Bookchin, Murray (1980), *Toward an Ecological Society.* Montreal: Black Rose.

Bookchin, Murray (1982), *The Ecology of Freedom: The Emergence and Dissolution of Hierarchy.* Palo Alto, Cal.: Cheshire.

Bornstein, Morris and Daniel R. Fusfield (eds) (1974), *The Soviet Economy* (4th edn). Homewood, Ill.: Richard Irwin.

Boulding, Kenneth E. (1966), "The Economics of Knowledge and the Knowledge of Economics." *American Economic Review, Papers and Proceedings* 56: 1–13.

Boulding, Kenneth E. (1977), "Commons and Community: The Idea of a Public," pp. 280–94 in Garrett Hardin and John Baden (eds), *Managing the Commons.* San Francisco: W.H. Freeman.

Brewer, Garry and Peter deLeon (1983), *Foundations of Policy Analysis*. Homewood, Ill.: Dorsey Press.

Britan, G. and B.S. Denich (1976), "Environment and Choice in Rapid Social Change." *American Ethnologist* 3: 55–72.

Brown, Lester (1974), *In the Human Interest: A Strategy to Stabilize World Population*. New York: W.W. Norton.

Brown, Lester R., et al. (1985), *State of the World 1985: A Worldwatch Institute Report on Progress Toward a Sustainable Society*. New York: W.W. Norton.

Bruner, Jerome S. (1962), "The Conditions of Creativity," pp. 1–30 in Howard E. Gruber, Glenn Terell, and Michael Wertheimer (eds), *Contemporary Approaches to Creative Thinking*. New York: Atherton.

Bull, Hedley (1977), *The Anarchical Society: A Study of Order in World Politics*. London: Macmillan.

Bullock, Kari and John Baden (1977), "Communes and the Logic of the Commons," pp. 182–99 in Garrett Hardin and John Baden (eds), *Managing the Commons*. San Francisco: W.H. Freeman.

Butzer, Karl W. (1980), "Civilizations: Organisms or Systems?" *American Scientist* 68: 517–23.

Cahn, Robert (1978), *Footprints on the Planet: A Search for an Environmental Ethic*. New York: Universe.

Caldwell, Lynton Keith (1984), *International Environmental Policy: Emergence and Dimensions*. Durham, NC: Duke University Press.

Campbell, Donald T. (1969), "Reforms as Experiments." *American Psychologist* 24: 409–29.

Campbell, Donald T. and Julian Stanley (1966), *Experimental and Quasi-Experimental Designs for Research*. Chicago: Rand McNally.

Capra, Fritjof (1982), *The Turning Point: Science, Society, and the Rising Culture*. New York: Simon and Schuster.

Catton, W.R. (1980), *Overshoot: The Ecological Basis of Revolutionary Change*. Urbana: University of Illinois Press.

Clark, Colin W. (1974), "The Economics of Overexploitation." *Science* 181: 630–4.

Clark, Margaret L. (1984), "Representative Technocracy: The Case of Antarctica." Mimeo, Ohio State University, Columbus.

Clark, Peter B. (1980), "Mediating Energy, Environmental, and Economic Conflict over Fuel Policy for Power Generation in New England," pp. 173–204 in Laura M. Lake (ed.), *Environmental Mediation: The Search for Consensus*. Boulder, Col.: Westview Press.

Coase, Ronald H. (1960), "The Problem of Social Cost." *Journal of Law and Economics* 3: 1–44.

Cohen, M.D., J.G. March, and J.P. Olson (1972), "A Garbage Can Model of Organizational Choice." *Administrative Science Quarterly* 17: 1–25.

Cohen, Steven (1984), "Defusing the Toxic Time Bomb: Federal Hazardous Waste Programs," pp. 273–92 in Norman J. Vig and Michael E. Kraft (eds), *Environmental Policy in the 1980s: Reagan's New Agenda*. Washington, DC: Congressional Quarterly Press.

Cole, Sam (1978), "The Global Futures Debate, 1965–1976," pp. 9–49 in Christopher Freeman and Marie Jahoda (eds), *World Futures: The Great Debate*. New York: Universe Books.

Colinvaux, Paul (1978), *Why Big Fierce Animals are Rare: An Ecologist's Perspective*. Princeton: Princeton University Press.

Colinvaux, Paul A. (1982), "Towards a Theory of History: Fitness, Niche, and Clutch of Homo Sapiens." *Journal of Ecology* 70: 393–412.

Commoner, Barry (1972), *The Closing Circle*. New York: Bantam Press.

Cone, John D. and Stephen C. Hayes (1982), *Environmental Problems/ Behavioral Solutions*. Belmont, Cal.: Brooks Cole.

Cormick, Gerald W. (1976), "Mediating Environmental Controversies: Perspectives and First Experience." *Earth Law Journal* 2: 215–24.

Cotgrove, Stephen (1982), *Catastrophe or Cornucopia: The Environment, Politics, and the Future*. New York: John Wiley.

Crecine, John P. (1982), "Information Processing Approaches to Political and Social Sciences." Mimeo, Ohio State University, Columbus.

Crenson, Matthew A. (1971), *The Un-Politics of Air Pollution*. Baltimore: Johns Hopkins University Press.

Crowe, Beryl L. (1969), "The Tragedy of the Commons Revisited." *Science* 166: 1103–7.

Culhane, Paul J. (1984), "Sagebrush Rebels in Office: Jim Watt's Land and Water Politics," pp. 293–317 in Norman J. Vig and Michael E. Kraft (eds), *Environmental Policy in the 1980s: Reagan's New Agenda*. Washington, DC: Congressional Quarterly Press.

Cyert, Richard and James G. March (1963), *A Behavioral Theory of the Firm*. Englewood Cliffs, NJ: Prentice-Hall.

Dahl, Robert A. (1971), *Polyarchy*. New Haven, Conn.: Yale University Press.

Dahl, Robert A. and Charles E. Lindblom (1953), *Politics, Economics, and Welfare: Planning and Politico-Economic Systems Resolved into Basic Social Processes*. Chicago: University of Chicago Press.

Dale, Tom and Vernon Gill Carter (1955), *Topsoil and Civilization*. Norman: Oklahoma University Press.

Dales, J.H. (1968), *Pollution, Property, and Prices*. Toronto: University of Toronto Press.

Daly, Herman E. (ed.) (1973), *Toward a Steady-State Economy*. San Francisco: W.H. Freeman.

Daly, Herman E. (1977), *Steady-State Economics*. San Francisco: W.H. Freeman.

Dawes, Robyn M. (1980), "Social Dilemma." *Annual Review of Psychology* 31: 169–93.

Dawes, R., J. McTavish, and H. Shaklee (1977), "Behavior, Communication, and Assumptions about Other Peoples' Behavior in a Commons Dilemma Situation." *Journal of Personality and Social Psychology* 35: 1–11.

Demsetz, Harold (1967), "Toward a Theory of Property Rights." *American Economic Review* 57: 347–59.

Deutsch, Karl W. (1977), "On the Interaction of Ecological and Political

Systems: Some Potential Contributions of the Social Sciences to the Study of Man and his Environment," pp. 23–31 in Karl W. Deutsch (ed.), *Ecosocial Systems and Ecopolitics*. Paris: UNESCO.

Dewey, John (1922), *Human Nature and Conduct*. New York: Modern Library.

Diesing, Paul (1962), *Reason in Society*. Urbana: University of Illinois Press.

Dolman, Anthony J. (1981), *Resources, Regimes, and World Order*. New York: Pergamon Press.

Domhoff, William G. (1978), *The Powers That Be: Processes of Ruling Class Domination in America*. New York: Random House.

Dror, Yehezkel (1964), "Muddling Through: Science or Inertia?" *Public Administration Review* 24: 153–65.

Dryzek, John (1982), "Policy Analysis as a Hermeneutic Activity." *Policy Sciences* 14: 309–29.

Dryzek, John S. (1983a), *Conflict and Choice in Resource Management: The Case of Alaska*. Boulder, Col.: Westview Press.

Dryzek, John S. (1983b), "Ecological Rationality." *International Journal of Environmental Studies* 21: 5–10.

Dryzek, John S. (1983c), "Present Choices, Future Consequences: A Case for Thinking Strategically." *World Futures* 19: 1–19.

Dryzek, John S. (1983d), "Don't Toss Coins in Garbage Cans: A Prologue to Policy Design." *Journal of Public Policy* 3: 345–68.

Dryzek, John S. and Amy S. Glenn (1987), "The Political Economy of Deforestation in Costa Rica." In Fred Hitzhusen and Robert MacGregor (eds), *A Multidisciplinary Approach to Third World Renewable Energy*. Columbus, Ohio: Horizons.

Dryzek, John S. and Robert E. Goodin (1986), "Risk-Sharing and Social Justice: The Motivational Foundations of the Post-War Welfare State." *British Journal of Political Science* 16: 1–34.

Dryzek, John S. and Susan Hunter (1987), "Environmental Mediation for International Problems." *International Studies Quarterly* 31.

Dryzek, John S. and Oran R. Young (1985), "Internal Colonialism in the Circumpolar North: The Case of Alaska." *Development and Change* 16: 123–45.

Dubos, Renee (1980), *The Wooing of the Earth*. New York: Charles Scribner's Sons.

Duncan, Otis D. (1964), "Social Organization and the Ecosystem," pp. 36–78 in R.E.L. Faris (ed.), *Handbook of Modern Sociology*. Chicago: Rand McNally.

Eckholm, Eric P. (1982), *Down to Earth: Environment and Human Needs*. London: Pluto Press.

Edelman, Murray (1977), *Political Language: Words that Succeed and Policies that Fail*. New York: Academic Press.

Edney, J.J. (1981), "Paradoxes on the Commons: Scarcity and the Problem of Inequality." *Journal of Community Psychology* 9: 3–34.

Edson, M.M., T.C. Foin, and C.M. Knapp (1981), "Emergent Properties and Ecological Research." *American Naturalist* 118: 593–6.

Ehrenfeld, David (1978), *The Arrogance of Humanism*. New York: Oxford University Press.

Ehrlich, Paul (1968),*The Population Bomb*. New York: Ballantine.

Elkin, Stephen L. (1983), "Economic and Political Rationality." Paper presented at the Midwest Political Science Association Convention.

Ellen, Roy (1982), *Environment, Subsistence, and System: The Ecology of Small-Scale Social Formations*. Cambridge, UK: Cambridge University Press.

Engelberg, J. and L.L. Boyarsky (1979), "The Noncybernetic Nature of Eco-systems." *American Naturalist* 114: 317–24.

Enloe, Cynthia (1975), *The Politics of Pollution in Comparative Perspective*. New York: David McKay.

Falk, Richard A. (1971), *This Endangered Planet: Prospects and Proposals for Human Survival*. New York: Random House.

Falk, Richard A. (1975), *A Study of Future Worlds*. New York: Free Press.

Falk, Richard A. (1983), *The End of World Order*. New York: Holmes and Meier.

Farer, Tom J. (1977), "The Greening of the Globe: A Preliminary Appraisal of the World Order Models Project." *International Organization* 31: 129–48.

Feyerabend, Paul (1978), *Science in a Free Society*. London: New Left Books.

Fisher, Roger and William Ury (1981), *Getting to Yes*. Boston: Houghton Mifflin.

Flathman, Richard E. (1976), *The Practice of Rights*. Cambridge, UK: Cambridge University Press.

Fried, Charles (1978), *Right and Wrong*. Cambridge, Mass.: Harvard University Press.

Friedman, James M. and Michael S. McMahon (1984), *The Silent Alliance: Canadian Support for Acid Rain Controls in the United States and the Campaign for Additional Electricity Exports*. Chicago: Regnery Gateway.

Friedman, Milton and Rose Friedman (1962), *Capitalism and Freedom*. Chicago: University of Chicago Press.

Fullenbach, Josef (1981), *European Environmental Policy: East and West* (trans. Frank Carter and John Manton). London: Butterworths.

Galbraith, John K. (1958), *The Affluent Society*. Boston: Houghton Mifflin.

Georgescu-Roegen, Nicholas (1971), *The Entropy Law and the Economic Process*. Cambridge, Mass.: Harvard University Press.

Georgescu-Roegen, Nicholas (1979), "Energy Analysis and Economic Valu-ation." *Southern Economic Journal* 45: 1023–58.

Gerth, H.H. and C. Wright Mills (1948), *From Max Weber: Essays in Sociology*. London: Routledge and Kegan Paul.

Ghiselin, Michael T. (1974), *The Economy of Nature and the Evolution of Sex*. Berkeley: University of California Press.

Goldman, Marshall I. (1972), *The Spoils of Progress: Environmental Pollution in the Soviet Union*. Cambridge, UK: Cambridge University Press.

Goldsmith, Edward, et al. (1972), *Blueprint for Survival*. Boston: Houghton Mifflin.

Goodin, Robert E. (1980), "Making Moral Incentives Pay." *Policy Sciences* 12: 131–45.

Goodin, Robert E. (1982), *Political Theory and Public Policy*. Chicago: University of Chicago Press.

Goodin, Robert E. and Ilmar Waldner (1979), "Thinking Big, Thinking Small, and Not Thinking At All." *Public Policy* 27: 1–24.

Gordon, H. Scott (1954), "The Economic Theory of a Common-Property Resource: The Fishery." *Journal of Political Economy* 62: 124–42.

Gorham, Eville (1982), "What To Do About Acid Rain." *Technology Review* 85 (7): 58–70.

Gorz, Andre (1980), *Ecology as Politics*. Boston: South End Press.

Grey, Thomas C. (1976), "Property and Need: The Welfare State and Theories of Distributive Justice." *Stanford Law Review* 28: 877–902.

Guess, George (1979), "Pasture Expansion, Forestry, and Development Contradictions: The Case of Costa Rica." *Studies in Comparative International Development* 14: 42–55.

Habermas, Jurgen (1970a), "On Systematically Distorted Communication." *Inquiry* 13: 205–18.

Habermas, Jurgen (1970b), "Toward a Theory of Communicative Competence." *Inquiry* 13: 360–75.

Habermas, Jurgen (1971), *Knowledge and Human Interests* (trans. Jeremy J. Shapiro). Boston: Beacon Press.

Habermas, Jurgen (1973a), *Legitimation Crisis*. Boston: Beacon Press.

Habermas, Jurgen (1973b), *Theory and Practice* (trans. John Viertel). Boston: Beacon Press.

Habermas, Jurgen (1984), *The Theory of Communicative Action* I: *Reason and the Rationalization of Society* (trans. Thomas McCarthy). Boston: Beacon Press.

Haefele, Edwin T. (1973), *Representative Government and Environmental Management*. Baltimore: Johns Hopkins University Press for Resources for the Future.

Hall, C.A.S. (1975), "The Biosphere, the Industriosphere, and their Interactions." *Bulletin of the Atomic Scientists* 31: 11–21.

Hardin, Garrett (1968), "The Tragedy of the Commons." *Science* 162: 1243–8.

Hardin, Garrett (1977), "Living on a Lifeboat," pp. 261–79 in Garrett Hardin and John Baden (eds), *Managing the Commons*. San Francisco: W.H. Freeman.

Hardin, Russell (1982), *Collective Action*. Baltimore: Johns Hopkins University Press for Resources for the Future.

Harman, Willis W. (1976), *An Incomplete Guide to the Future*. New York: W.W. Norton.

Harris, Marvin (1966), "The Cultural Ecology of India's Sacred Cattle." *Current Anthropologist* 7: 51–9.

Harsanyi, J.C. (1975), "Can the Maximin Principle Serve as a Basis for Morality?" *American Political Science Review* 69: 594–606.

Hart, H.L.A. (1955), "Are There Any Natural Rights?" *Philosophical Review* 64: 175–91.

Hart, H.L.A. (1958), "Positivism and the Separation of Law and Morals." *Harvard Law Review* 71: 593–629.

Hart, H.L.A. (1961), *The Concept of Law*. Oxford: Clarendon Press.

Hawken, Paul, James Ogilvy, and Peter Schwartz (1982), *Seven Tomorrows: Toward a Voluntary History*. New York: Bantam Press.

Hawley, A.H. (1950), *Human Ecology: A Theory of Community Structure*. New York: Ronald Press.

Hayek, Friedrich A. von (1979), *Law, Legislation, and Liberty: The Political Order of a Free People*. Chicago: University of Chicago Press.

Hayek, Friedrich A. von (1979), *Law, Legislation, and Liberty: The Political Order of a Free People*. Chicago: University of Chicago Press, 1979.

Head, John (1974), *Public Goods and Public Welfare*. Durham, NC: Duke University Press.

Heilbroner, Robert L. (1980), *An Inquiry Into the Human Prospect: Updated and Reconsidered for the 1980s*. New York: W.W. Norton.

Henderson, Hazel (1981), *The Politics of the Solar Age: Alternatives to Economics*. Garden City, NY: Anchor Books.

Hess, Karl (1979), *Community Technology*. New York: Harper and Row.

Hirshleifer, J. (1977), "Economics from a Biological Viewpoint." *Journal of Law and Economics* 20: 1–52.

Hobbes, Thomas (1946), *Leviathan* (ed. Michael Oakeshott). Oxford: Basil Blackwell.

Hofstadter, Douglas R. (1979), *Gödel, Escher, Bach: An Eternal Golden Brain*. New York: Basic Books.

Horkheimer, Max (1947), *Eclipse of Reason*. New York: Oxford University Press.

Hume, David (1777), *An Enquiry Concerning the Principles of Morals*. London: T. Cadell.

Hummel, Ralph P. (1982), *The Bureaucratic Experience* (2nd edn). New York: St Martin's Press.

Inbar, Michael (1979), *Routine Decision-Making: The Future of Bureaucracy*. Beverly Hills: Sage.

Jaffe, Louis L. and Laurence H. Tribe (1971), *Environmental Protection*. Chicago: Bracton.

James, Roger (1980), *Return to Reason: Popper's Thought in Public Life*. Shepton Mallet, Somerset, UK: Open Books.

Jenny, Hans (1980), "Alcohol or Humus?" *Science* 209: 444.

Jerdee, T. and B. Rosen (1974), "Effects of Opportunity to Communicate and Visibility of Individual Decisions on Behavior in the Common Interest." *Journal of Applied Psychology* 5: 712–16.

Johnson, Warren (1979), *Muddling Toward Frugality*. Boulder, Col.: Shambhala Press.

Jones, Charles O. (1975), *Clean Air: The Policies and Politics of Pollution Control*. Pittsburgh: University of Pittsburgh Press.

Kahn, Herman, et al. (1976), *The Next Two Hundred Years*. New York: William Morrow.

Kapitsa, Pyotr, et al. (1977), *Society and the Environment: A Soviet View* (trans. John Williams). Moscow: Progress Publishers.

Kapp, K. William (1971), *The Social Costs of Private Enterprise*. New York: Shocken.

Kaufman, Herbert (1960), *The Forest Ranger: A Study in Administrative Behavior*. Baltimore: Johns Hopkins University Press for Resources for the Future.

Kelley, Donald R. (1976), "Environmental Policy Making in the USSR." *Soviet Studies* 28: 570–89.

Kelley, Donald R., Kenneth R. Stunkel, and Richard R. Wescott (1976), *The Economic Superpowers and the Environment*. San Francisco: W.H. Freeman.

Keohane, Robert O. and Joseph S. Nye (1977), *Power and Interdependence: World Politics in Transition*. Boston: Little Brown.

Keynes, J.M. (1936), *The General Theory of Employment, Interest, and Money*. London: Macmillan.

Kothari, Rajni (1974), *Footsteps into the Future: Diagnosis of the Present World and a Design for an Alternative*. New Delhi: Orient Longman.

Kuhn, Thomas S. (1962), *The Structure of Scientific Revolutions*. Chicago: University of Chicago Press.

Kuznets, Simon (1966), *Modern Economic Growth*. New Haven, Conn.: Yale University Press.

Lakatos, Imre (1970), "Falsification and the Methodology of Scientific Research Programmes," pp. 91–196 in Imre Lakatos and Alan Musgrave (eds), *Criticism and the Growth of Knowledge*. Cambridge: Cambridge University Press.

Lake, Laura M. (ed.) (1980), *Environmental Mediation: The Search for Consensus*. Boulder, Col.: Westview Press.

Lane, Robert E. (1978), "Autonomy, Felicity, Futility: The Effects of the Market Economy on Political Personality." *Journal of Politics* 40: 2–24.

LaPorte, Todd R. (1975), "Complexity and Uncertainty: Challenge to Action," pp. 332–56 in Todd R. LaPorte (ed.), *Organized Social Complexity: Challenges to Politics and Policy*. Princeton: Princeton University Press.

Lasswell, Harold D. (1965), "The World Revolution of our Time: A Framework for Basic Research," pp. 29–96 in Harold D. Lasswell and Daniel Lerner (eds), *World Revolutionary Elites*. Cambridge, Mass.: MIT Press.

Laudan, Larry (1977), *Progress and its Problems: Toward a Theory of Scientific Growth*. Berkeley: University of California Press.

Le Carré, John (1965), *The Looking-Glass War*. New York: Coward, McCann, and Geoghegan.

Lemons, J. (1981), "Cooperation and Stability as a Basis for Environmental Ethics." *Environmental Ethics* 3: 219–30.

Leopold, Aldo (1933), *Game Management*. New York: Charles Scribner's Sons.

Leopold, Aldo (1949), *A Sand County Almanac*. Oxford: Oxford University Press.

Levine, Robert A. (1972), *Public Planning: Failure and Redirection*. New York: Basic Books.

Lindblom, Charles E. (1959), "The Science of Muddling Through." *Public Administration Review* 19: 79–88.

Lindblom, Charles E. (1965), *The Intelligence of Democracy: Decision Making Through Mutual Adjustment*. New York: Free Press.

Lindblom, Charles E. (1977), *Politics and Markets: The World's Political-Economic Systems*. New York: Basic Books.

Lindblom, Charles E. (1982), "The Market as Prison." *Journal of Politics* 44: 324–36.

Lindblom, Charles E. and David K. Cohen (1979), *Usable Knowledge: Social Science and Social Problem Solving*. New Haven, Conn.: Yale University Press.

Lovins, Amory B. (1977), *Soft Energy Paths: Toward a Durable Peace*. New York: Harper and Row.

Lowi, Theodore J. (1969), *The End of Liberalism*. New York: W.W. Norton.

Luke, Timothy W. and Stephen K. White (1985), "Critical Theory, the Informational Revolution, and an Ecological Modernity." In John Forester (ed.), *Critical Theory and Public Life*. Cambridge, Mass.: MIT Press.

MacCallum, Gerald C. (1968), "Legislative Intent," pp. 237–73 in Robert S. Summers (ed.), *Essays in Legal Philosophy*. Berkeley: University of California Press.

MacIntyre, Alasdair (1981), *After Virtue*. Notre Dame, Ind.: University of Notre Dame Press.

McCarthy, Thomas (1978), *The Critical Theory of Jurgen Habermas*. Cambridge, Mass.: MIT Press.

McCay, Bonnie J. (1978), "Systems Ecology, People Ecology, and the Anthropology of Fishing Communities." *Human Ecology* 6: 397–421.

McClosky, H.J. (1983), *Ecological Ethics and Politics*. Totowa, NJ: Rowan and Littlefield.

McFarland, Andrew (1984), "An Experiment in Regulatory Negotiation: The National Coal Policy Project." Paper presented at the Annual Meeting of the Western Political Science Association.

McKean, Roland N. (1968), "The Use of Shadow Prices." In S.B. Chase (ed.), *Problems in Public Expenditure Analysis*. Washington, DC: Brookings Institution.

Magee, Brian (1973), *Popper*. London: Fontana.

Mannheim, Karl (1940), *Man and Society in an Age of Reconstruction*. London: Kegan Paul.

Margalef, Ramon (1973), "Ecological Theory and Prediction in the Study of Interaction Between Man and the Rest of the Biosphere," pp. 307–53 in Harold Sioli (ed.), *Ekologie und Lebensschutz in Internationaler Sicht*. Freiburg: Verlag Rombach.

Martin, John R. (1979), "The Concept of the Irreplaceable." *Environmental Ethics* 1: 31–48.

Martindale, Don (1960), *The Nature and Types of Sociology Theory*. Boston: Houghton Mifflin.

Maxwell, Robert J. (1981), *Health and Wealth*. Lexington, Mass.: Lexington Books.

Maynard Smith, John (1982), *Evolution and the Theory of Games*. Cambridge, UK: Cambridge University Press.

Meadows, Donella H., et al. (1972), *The Limits to Growth*. New York: Universe Books.

Mendlovitz, Saul H. (ed.) (1975), *On the Creation of a Just World Order: Preferred Worlds for the 1980s.* New York: Free Press.

Mernitz, Scott (1980), *Mediation of Environmental Disputes: A Sourcebook.* New York: Praeger.

Michelson, William (1970), *Man and His Urban Environment: A Sociological Approach.* Reading, Mass.: Addison-Wesley.

Milbrath, Lester W. (1984), *Environmentalists: Vanguard for a New Society.* Albany, NY: State University of New York Press.

Miles, Rufus E. Jr (1976), *Awakening from the American Dream: The Social and Political Limits to Growth.* New York: Universe Books.

Mills, C. Wright (1956), *The Power Elite.* New York: Oxford University Press.

Mitroff, Ian and L. Vaughan Blankenship (1973), "On the Methodology of the Holistic Experiment: An Approach to the Conceptualization of Large-Scale Social Experiments." *Technological Forecasting and Social Change* 4: 339–53.

Morgenthau, Hans (1967), *Politics Among Nations* (4th edn). New York: Alfred A. Knopf.

Mosher, Lawrence (1983), "Distrust of Gorsuch May Stymie EPA Attempt to Integrate Pollution Wars." *National Journal* 15: 322–4

Mueller, Dennis C. (1979), *Public Choice.* Cambridge, UK: Cambridge University Press.

Murray, Francis X. (ed.) (1978), *Where We Agree: Report of the National Coal Policy Project.* Boulder, Col.: Westview Press.

Myers, Norman (1980), *Conversion of Tropical Moist Forests.* Washington, DC: National Academy of Sciences.

Nef, John U. (1977), "An Early Energy Crisis and its Consequences." *Scientific American* 237: 140–51.

Nelson, Richard R. (1977), *The Moon and the Ghetto: An Essay on Public Policy Analysis.* New York: W.W. Norton

Nelson, R.K. (1982), "A Conservation Ethic and Environment: The Koyukon of Alaska," pp. 211–28 in N.M. Williams and E.S. Hunn (eds), *Resource Managers: North American and Australian Hunter-Gatherers.* Boulder, Col.: Westview Press.

Niskanen, William A. Jr (1971), *Bureaucracy and Representative Government.* Chicago: Aldine-Atherton.

Nozick, Robert (1974), *Anarchy, State, and Utopia.* New York: Basic Books.

Odum, Eugene P. (1983), *Basic Ecology.* Philadelphia: Saunders.

Olson, Mancur (1965), *The Logic of Collective Action.* Cambridge, Mass.: Harvard University Press.

Olson, Mancur (1982), *The Rise and Decline of Nations: Economic Growth, Stagflation, and Social Rigidities.* New Haven, Conn.: Yale University Press.

O'Neill, Gerard K. (1978), *The High Frontier.* New York: Bantam Books.

Ophuls, William (1977), *Ecology and the Politics of Scarcity.* San Francisco: W.H. Freeman.

Orlove, B.S. (1980), "Ecological Anthropology." *Annual Review of Anthropology* 9: 235–73.

Owen, D.F. (1973), *Man in Tropical Africa.* New York: Oxford University Press.

Page, Talbot (1977), *Conservation and Economic Efficiency*. Baltimore: Johns Hopkins University Press for Resources for the Future.

Parenti, Michael (1983), *Democracy for the Few* (4th edn). New York: St Martin's Press.

Patten, Bernard C. and Eugene P. Odum (1981), "The Cybernetic Nature of Ecosystems." *American Naturalist* 118: 886–95.

Pejovich, Svetozar (1972), "Toward an Economic Theory of the Creation and Specification of Property Rights." *Review of Social Economy* 30: 309–25.

Piven, Frances Fox and Richard A. Cloward (1972), *Regulating the Poor*. New York: Vintage.

Polanyi, Michael (1944), *The Great Transformation*. Boston: Beacon Press.

Popper, Karl (1961), *The Poverty of Historicism*. London: Routledge and Kegan Paul.

Popper, Karl (1966), *The Open Society and its Enemies*. London: Routledge and Kegan Paul.

Popper, Karl (1972), *Objective Knowledge*. Oxford: Clarendon Press.

Postel, Sandra (1985), "Protecting Forests from Air Pollution and Acid Rain," pp. 97–123 in Lester R. Brown, et al., *State of the World, 1985*. New York: W.W. Norton.

Pressman, Jeffrey and Aaron Wildavsky (1973), *Implementation*. Berkeley: University of California Press.

Pryde, Philip R. (1972), "The Quest for Environmental Quality in the USSR." *American Scientist* 60: 739–45.

Pugh, George Edgin (1977), *The Biological Origin of Human Values*. New York: Basic Books.

Ralston, Holmes B. III (1981), "Values in Nature." *Environmental Ethics* 3: 113–28.

Randall, Alan (1974), "Coase Externality Theory in a Policy Context." *Natural Resources Journal* 14: 34–54.

Randers, Jorgen and Donella Meadows (1973), "The Carrying Capacity of our Global Environment: A Look at the Ethical Alternatives," pp. 253–76 in Ian G. Barbour (ed.), *Western Man and Environmental Ethics*. Reading, Mass.: Addison-Wesley.

Rawls, John (1971), *A Theory of Justice*. Cambridge, Mass.: Harvard University Press.

Rein, Martin (1976), *Social Science and Public Policy*. Harmondsworth, Middx: Penguin.

Richerson, Peter J. (1977), Ecology and Human Ecology: A Comparison of Theories in the Biological and Social Sciences." *American Ethnologist* 4: 1–26.

Riddick, W.L. (1971), *Charrette Processes: A Tool in Urban Planning*. York: George Shumway.

Ridker, R.G. and J.A. Henning (1971), "The Determinants of Residential Property Values with Special Reference to Air Pollution." *Review of Economics and Statistics* 53: 246–57.

Rifkin, Jeremy (1981), *Entropy: A New World View*. New York: Bantam Books.

Rivlin, Alice (1973), "Forensic Social Science." In *Perspectives on Inequality*. Cambridge, Mass.: Harvard Educational Review Reprint Series, no. 8.

Roberts, Simon (1979), *Order and Dispute: An Introduction to Legal Anthropology*. Harmondsworth, Middx: Penguin.

Rollin, Bernard E. (1981), *Animal Rights and Human Morality*. Buffalo, NY: Prometheus.

Rosenberg, Nathan (1973), "Innovative Responses to Materials Shortages." *American Economic Review* 63: 111–18.

Rosenblum, Victor (1974), "The Continuing Role of the Courts in Allocating Common Property Resources," pp. 119–41 in Edwin T. Haefele (ed.), *The Governance of Common Property Resources*. Baltimore: Johns Hopkins University Press for Resources for the Future.

Rousseau, Jean-Jacques (1968), *The Social Contract* (trans. and introduced by Maurice Cranston). Harmondsworth, Middx: Penguin.

Rousseau, Jean-Jacques (1972), *The Government of Poland* (trans. Willmoore Kendall). Indianapolis: Bobbs-Merrill.

Ruggie, John Gerard (1975), "Complexity, Planning, and Public Order," pp. 119–50 in Todd R. LaPorte (ed.), *Organized Social Complexity: Challenge to Politics and Policy*. Princeton: Princeton University Press.

Sagoff, Mark, (1974), "On Preserving the Natural Environment." *Yale Law Journal* 84: 205–67.

Sahlins, Marshall (1976), *Culture and Practical Reason*. Chicago: University of Chicago Press.

Sale, Kirkpatrick (1980), *Human scale*. New York: G.P. Putnam's Sons.

Sax, Joseph L. (1971), *Defending the Environment: A Strategy for Citizen Action*. New York: Alfred A. Knopf.

Schell, Jonathan (1982), *The Fate of the Earth*. New York: Alfred A. Knopf.

Schelling, Thomas C. (1960), *The Strategy of Conflict*. Cambridge, Mass.: Harvard University Press.

Schick, Allen (1973), "A Death in the Bureaucracy: The Demise of Federal PPB." *Public Administration Review* 33: 146–56.

Schmitter, Phillippe C. and Gerhard Lehmbruch (eds) (1979), *Trends Toward Corporatist Intermediation*. Beverly Hills: Sage.

Schumacher, E.F. (1973), *Small is Beautiful: Economics as if People Mattered*. New York: Harper and Row.

Schumacher, E.F. (1979), *Good Work*. New York: Harper and Row.

Schurmann, Herbert F. (1968), *Ideology and Organization in Communist China*. Berkeley: University of California Press.

Scott, Anthony (1973), *Natural Resources: The Economics of Conservation*. Toronto: McClelland Stewart.

Sears, Paul B. (1957), *The Ecology of Man*. Corvallis: Oregon State University Press.

Sewell, W.R. Derrick and Timothy O'Riordan (1976), "The Culture of Participation in Environmental Decision-Making." In Alfred E. Utton, W.R. Derrick Sewell, and Timothy O'Riordan (eds), *Natural Resources for a Democratic Society: Public Participation in Decision-Making*. Boulder, Col.: Westview Press.

Shiva, Vandana, H.C. Sharatchandra, and J. Bandyopadhyay (1982), "Social Forestry – No Solution Within the Market." *The Ecologist* 12: 158–68.

Shue, Henry (1980), *Basic Rights: Subsistence, Affluence, and US Foreign Policy.* Princeton : Princeton University Press.

Sibley, Mulford Q. (1977), *Nature and Civilization: Some Implications for Politics.* Ithaca, NY: F.E. Peacock.

Simon, Herbert A. (1981), *The Sciences of the Artificial* (2nd edn). Cambridge, Mass.: MIT Press.

Simon, Julian (1981), *The Ultimate Resource.* Princeton: Princeton University Press.

Simon, Julian L. and Herman Kahn (eds) (1984), *The Resourceful Earth.* Oxford: Basil Blackwell.

Singleton, Fred (ed.) (1976), *Environmental Misuse in the Soviet Union.* New York: Praeger.

Smil, Vaclav (1984), *The Bad Earth: Environmental Degradation in China.* New York: M.E. Sharpe.

Smith, V. Kerry (ed.) (1979), *Scarcity and Growth Reconsidered.* Baltimore: Johns Hopkins University Press for Resources for the Future.

Spurlock, J.M. and M. Modell (1978), "Technology Requirements and Planning Criteria for Closed Life Support Systems for Manned Space Missions." Washington DC: NASA.

Stein, Arthur A. (1983), "Coordination and Collaboration: Regimes in an Anarchic World," pp. 115–40 in Stephen D. Krasner (ed.), *International Regimes.* Ithaca, NY: Cornell University Press.

Steinbruner, John D. (1974), *The Cybernetic Theory of Decision: New Dimensions of Political Analysis.* Princeton: Princeton University Press.

Stone, Christopher D. (1972), "Should Trees Have Standing? Toward Legal Rights for Natural Objects." *Southern California Law Review* 45: 450–501.

Strange, Susan (1983), "*Cave! Hic Dragonis*: A Critique of Regime Analysis," pp. 337–54 in Stephen D. Krasner (ed.), *International Regimes.* Ithaca, NY: Cornell University Press.

Stretton, H. (1976), *Capitalism, Socialism, and the Environment.* Cambridge, UK: Cambridge University Press.

Stroup, Richard and John Baden (1983), *Natural Resources: Bureaucratic Myths and Environmental Management.* Cambridge, Mass.: Ballinger Press.

Talbot, Allan R. (1983), *Settling Things: Six Case Studies in Environmental Mediation.* Washington, DC: Conservation Foundation.

Taylor, Michael (1976), *Anarchy and Cooperation.* London: John Wiley.

Taylor, Michael (1982), *Community, Anarchy, and Liberty.* Cambridge, UK: Cambridge University Press.

Thayer, Frederick C. (1981), *An End to Hierarchy and Competition.* New York: New Viewpoints.

Thurow, Lester C. (1980), *The Zero-Sum Society: Distribution and the Possibilities for Economic Change.* New York: Basic Books.

Titmuss, Richard M. (1971), *The Gift Relationship: From Human Blood to Social Policy.* New York: Pantheon Books.

Toffler, Alvin (1981). *The Third Wave*. New York: Bantam Books.

Touval, Saadia (1982), *The Peace Brokers: Mediators in the Arab–Israeli Conflict, 1948–1979*. Princeton: Princeton University Press.

Tribe, Laurence H. (1972), "Technology Assessment and the Fourth Discontinuity: The Limits of Instrumental Rationality." *Southern California Law Review* 46: 617–60.

Tribe, Laurence H. (1974), "Ways Not to Think About Plastic Trees: New Foundations for Environmental Law." *Yale Law Journal* 83: 1315–48.

Tucker, Robert C. (1967), "The De-radicalization of Marxist Movements." *American Political Science Review* 61: 343–58.

Unger, Roberto (1975), *Knowledge and Politics*. New York: Free Press.

Van Voris, P., R.V. O'Neill, W.R. Emanuel, and H.H. Shugart (1980), "Functional Complexity and Ecosystem Stability." *Ecology* 61: 1352–60.

Vernadsky, V.I. (1945), "The Biosphere and the Noosphere." *American Scientist* 33: 1–12.

Vig, Norman J. and Michael E. Kraft (eds) (1984), *Environmental Policy in the 1980s: Reagan's New Agenda*. Washington, DC: Congressional Quarterly Press.

Walker, Charles A. (1983), "Science and Technology of the Sources and Management of Radioactive Wastes," pp. 27–74 in Charles A. Walker, Leroy C. Gould and Edward J. Woodhouse (eds), *Too Hot to Handle? Social and Policy Issues in the Management of Radioactive Wastes*. New Haven, Conn.: Yale University Press.

Wall, James A. Jr (1981), "Mediation: An Analysis, Review, and Proposed Research." *Journal of Conflict Resolution* 25: 157–80.

Ward, Benjamin (1972), *What's Wrong with Economics?* New York: Basic Books.

Ward, Colin (1977), "Topless Federations," pp. 319–25 in George Woodcock (ed.), *The Anarchist Reader*. London: Fontana.

Weaver, Warren (1948), "Science and Complexity." *American Scientist* 36: 536–44.

Weber, Max (1968), "Bureaucracy," pp. 956–1005 in Max Weber, *Economy and Society* (ed. Guenther Roth and Klaus Wittich). New York: Bedminster Press.

Weidenbaum, Murray L. (1983), "Free the Fortune 500," pp. 72–9 in Theodore D. Goldfarb (ed.), *Taking Sides: Clashing Views on Controversial Environmental Issues*. Guildford: Dushkin.

Weinberg, Alvin M. (1972), "Social Institutions and Nuclear Energy." *Science* 177: 27–34.

Wellmer, Albrecht (1985), "Reason, Utopia, and the *Dialectics of Enlightenment*," pp. 35–66 in Richard J. Bernstein (ed.), *Habermas and Modernity*. Cambridge, UK: Polity Press.

Wenner, Lettie McSpadden (1976), *One Environment Under Law*. Pacific Palisades, Cal.: Goodyear.

Werbos, Paul (1979), "Changes in Policy Analysis Procedures Suggested by New Optimization Methods and Required for Global Survival." *Policy Analysis and Information Systems* 3: 27–52.

Wessel, Milton R. (1976), *The Rule of Reason: A New Approach to Corporate Litigation*. Reading, Mass.: Addison-Wesley.

Wetstone, Gregory S. and Armin Rosencranz (1983), *Acid Rain in Europe and North America*. Washington, DC: Environmental Law Institute.

White, Lynn Jr (1967), "The Historical Roots of Our Ecologic Crisis." *Science* 155: 1203–7.

Wildavsky, Aaron (1966), "The Political Economy of Efficiency." *Public Administration Review* 26: 292–310.

Wildavsky, Aaron (1979), *The Politics of the Budgetary Process* (3rd edn). Boston: Little Brown.

Wilensky, Harold L. (1967), *Organizational Intelligence*. New York: Basic Books.

Williams, Bernard (1979), "Conflicts of Values," pp. 221–32 in Alan Ryan (ed.), *The Idea of Freedom: Essays in Honour of Isaiah Berlin*. Oxford: Oxford University Press.

Wilson, David Sloan (1980), *The Natural Selection of Populations and Communities*. Menlo Park: Benjamin/Cummings.

Wilson, E.O. (1975), *Sociobiology: The New Synthesis*. Cambridge Mass.: Harvard University Press.

Wilson, Edward O. (1978), *On Human Nature*. Cambridge, Mass: Harvard University Press.

Winner, Langdon (1975), "Complexity and the Limits of Human Understanding," pp. 40–76 in Todd R. LaPorte (ed.), *Organized Social Complexity: Challenges to Politics and Policy*. Princeton: Princeton University Press.

Wolfe, Tom (1981), *From Bauhaus to Our House*. New York: Farrar, Straus, and Giroux.

Woodcock, George (1977), "Anarchism: An Historical Introduction," pp. 11–56 in George Woodcock (ed.), *The Anarchist Reader*. London: Fontana.

Woodhouse, Edward J. (1983), "The Politics of Nuclear Waste Management," pp. 151–83 in Charles A. Walker, Leroy C. Gould, and Edward J. Woodhouse (eds), *Too Hot to Handle? Social and Policy Issues in the Management of Radioactive Wastes*. New Haven, Conn.: Yale University Press.

Young, Oran R. (1967), *The Intermediaries: Third Parties in International Crises*. Princeton: Princeton University Press.

Young, Oran R. (1972), "The Actors in World Politics," pp. 125–44 in James Rosenau, Vincent Davis, and Maurice East (eds), *The Analysis of International Politics*. New York: Free Press.

Young, Oran R. (ed.) (1975), *Bargaining: Formal Theories of Negotiation*. Urbana, Ill.: University of Illinois Press.

Young, Oran R. (1978a), "Anarchy and Social Choice: Reflections on the International Polity." *World Politics* 30: 241–63.

Young, Oran R. (1978b), "On the Performance of the International Polity." *British Journal of International Studies* 4: 191–208.

Young, Oran R. (1979), *Compliance and Public Authority: A Theory with International Applications*. Baltimore: Jonhs Hopkins University Press for Resources for the Future.

Young, Oran R. (1980), "International Regimes: Problems of Concept Formation." *World Politics* 32: 331–56.

Young, Oran R. (1982), *Resource Regimes: Natural Resources and Social Institutions.* Berkeley: University of California Press, 1982.

Zile, Zigurds L. (1982), "Glimpses of the Scientific Revolution in Soviet Environmental Law," pp. 187–216 in Peter B. Maggs, Gordon B. Smith, and George Ginsburg (eds), *Law and Economic Development in the Soviet Union.* Boulder, Col.: Westview Press.

Index

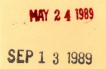